HIGH ENERGY PHYSICS AND APPLICATIONS

HIGH ENERGY PHYSICS AND APPLICATIONS

Proceedings of the UAE-CERN Workshop

Al Ain, United Arab Emirates
26 - 28 November 2007

EDITORS

John Ellis
CERN, Geneva, Switzerland

Salah Nasri
Ehab Malkawi
United Arab Emirates University
Al-Ain, United Arab Emirates

SPONSORING ORGANIZATIONS

Abu Dhabi Water & Electricity Authority
UAE National Energy & Water Research Center (NEWRC)
Microsoft Research
City Properties Real Estate
UAE Department of Health & Medical Services
COMSTECH
Abu Dhabi Food Control Authority
Dubai Silicon Oasis Authority
Pearson Education

AMERICAN INSTITUTE OF PHYSICS

Melville, New York, 2008

AIP CONFERENCE PROCEEDINGS ■ VOLUME 1006

Editors:

John Ellis
CERN - Theory Division
CH-1211 Geneva 23
Switzerland

E-mail: john.ellis@cern.ch

Salah Nasri
Ehab Malkawi

Department of Physics
United Arab Emirates University
P. O. Box 17551, Al-Ain
United Arab Emirates

E-mail: snasri@uaeu.ac.ae
emalkawi@uaeu.ac.ae

L.C. Catalog Card No. 2008924920
ISBN 978-0-7354-0525-7
ISSN 0094-243X
Printed in the United States of America

CONTENTS

PARALLEL SESSION CONTRIBUTIONS

Session A

Session B

POSTER SESSION

The European Laboratory for Particle Physics, CERN, is currently putting the finishing touches to the Large Hadron Collider (LHC) and its related experiments. The unparalleled energy and collision rates supplied by the LHC will provide new insights into the nature of matter and the contents of the Universe: its experiments will probe deeper inside matter than any previous particle accelerator, providing understanding of processes that occurred very early in the history of the Big Bang. The LHC will be not only the most powerful microscope ever built, but also an unequalled telescope. The LHC and its experiments continue humanity's longstanding efforts to understand better the Universe around us, inspired by the pioneering studies of Greek philosophers, Arab astronomers and Indian mathematicians, among many others.

In view of its universal interest, the LHC project has attracted unprecedented interest from physicists around the world, who are participating in increasing numbers. At the time of writing, the total number of scientists and engineers registered as participating in research activities at CERN exceeds 9000, who are based at universities and research institutes in dozens of countries around the world. Participation in research at CERN provides unique opportunities for students to advance their understanding of fundamental physics as well as make international contacts and master advanced technologies.

Advanced basic research requires the development and deployment of advanced technologies, and CERN and its partners around the world have been active in perfecting technological solutions in fields as diverse as cryogenic engineering, medical applications of accelerators and particle detectors, and informatics. Among the best-known innovations made at CERN has been the World-Wide Web, invented to enable scientists around the world to share data. CERN and its partners are now deploying the World-Wide LHC Computing Grid, which will enable physicists working on the LHC to share all their computing resources, including computers, storage media and software, as well as their experimental data. Distributed Grid computing technology has many applications to other sciences, ranging from environmental monitoring to computational chemistry, as well as other fields such as finance and the media.

For these scientific and technological reasons, in recent years there has been growing interest in collaboration between countries in the region of South Asia, the Near and Middle East, North Africa and CERN. Notable examples include Turkey, India, Morocco, Pakistan, Iran, and more recently the United Arab Emirates, Saudi Arabia, Egypt and Algeria. Several of these countries already participate in the LHC experiments, and some have also got involved in the construction of the LHC and the lower-energy accelerators that feed it with particles. Their physicists participate in fundamental research at the cutting edge, their students acquire unique training, and their institutes and industries gain lasting benefits from the novel technologies transferred from CERN. The United Arab Emirates, in particular, has for several years been sending physicists and students to CERN, and is actively interested in medical applications of particle detector technologies, as well as fundamental research and student training.

This workshop, held at Al Ain in the United Arab Emirates in November 2007, provided a unique opportunity for scientists from Europe and the region to meet each other, exchange their experiences, review the LHC programme, and discuss the possibilities for future collaboration. Many new contacts were made, between physicists from different countries in the region as well as with scientists from CERN and its European Member States. The latter particularly appreciated the opportunity to learn more about scientific activities in the region, as well as the status of their collaborations with CERN. The meeting also provided a unique possibility to discuss possible collaborations between countries in the region. All these scientific aspects of the meeting are reflected in contributions to these proceedings.

Visitors to Al Ain also appreciated the opportunity to learn more of the present-day achievements of the United Arab Emirates, as well as its history, its traditions and its generous hospitality. The economic progress being made in the region is truly impressive. We thank our hosts most sincerely for their warmth, thoughtfulness and kindness. We hope that, in the future, our collaboration will continue to grow for the benefits of all partners, for the advancement of science, and for the deepening of understanding between CERN, its Member States, and countries from the region represented at this meeting.

On behalf of all the participants, particularly those from outside the region, I should like to express our warm appreciation to our hosts in the United Arab Emirates, whose efforts made this meeting such a success.

John Ellis,
Advisor to the CERN Director-General for Relations with Non-Member States

Preface

Following the signature of an agreement between the United Arab Emirates and the European Organization for Nuclear Research (CERN) to promote scientific collaboration in the field of High Energy Physics and related technological applications to other fields of Science and Technology, the Department of Physics at the United Arab Emirates University, in close collaboration with CERN, organized a **three-day Workshop, from 26 to 28 November 2007**, under the title:"UAE-CERN Workshop: High Energy and Applications". The workshop aimed at:

- First, presenting CERN to the region: past, present and future research activities. Essentially those related to Elementary Particle Physics, but also those with impact on technological advances in areas outside the realm of high energy.
- Perhaps more importantly, creating an interactive medium where scientists in the field of High Energy Physics and related specialties meet, present their work, and engage in discussions.
- Equally important, exploring and engaging in ways in which collaborations can be fostered.

The topics presented at the workshop encompassed the following diverse subjects:

- All sub-fields of High Energy Physics: experimental, phenomenological, and theoretical. These include, but are not at all limited to:
  - SM Higgs search with CMS and ATLAS experiments.
  - Perspectives for SUSY and extra-dimensions searches at the LHC.
  - CP violation and B-Physics at LHCb.
  - Heavy-Ion Physics with ALICE experiment.
- Grid computing at CERN.

The workshop gathered people from many countries in the region, Europe, the U.S.A., and Japan.

We, the organizing committee, felt from all the participants that the workshop was very well organized and reached its goals successfully. This has created an enthusiasm and a momentum for more meetings to be organized in the regions. Ideas, such as a winter school in high energy physics organized in the region, among other excellent ideas, have sprung during the discussions held in the workshop. We believe that such ideas should be pursued and more meeting opportunities for high-energy physicists in the region should be organized more frequently. Our appreciations go the sponsors of the workshop and to all people who have made this workshop a success.

Organizing Committee

Local Organizing Committee

- Abdessamad Abada
- Noureddine Amrane
- Maamar Benkraouda
- Ilias Fernini
- Amal Al-Ghaferi
- Hassan Ghamlouche
- Doaa Ghoneim
- Fathalla Hamed
- Adel Hashish
- Mohamed Haj-Youssef
- Bashar Issa
- Jamila Al-Kaabi (Secretary)
- Usama Al-Khawaja
- Ehab Malkawi
- Sherif Moussa
- Salah Nasri
- Naser Qamhieh
- Saleh Thaker
- Hamda Al-Thani
- Nacir Tit (Committee Chair)
- Mahfood Bin Tursh (Relations)
- Mofreh Zaghloul

International Advisory Board

- Engin Arik (Istanbul, Turkey)
- Diether Blechschmidt (CERN)
- John Ellis (CERN)
- Hafeez Hoorani (Islamabad, Pakistan)
- Abdeslam Hoummada (Casablanca, Morocco)
- Panos Razis (Nicosia, Cyprus)
- Albert de Roeck (CERN)
- Turki Al-Saud (Riyadh, Saudi Arabia)
- Mohamed Sherif (Cairo, Egypt)
- EmmanuelTsesmelis(CERN)

Sponsors

Abu Dhabi Water & Electricity Authority

UAE National Energy & Water Research Center (NEWRC)

Microsoft Research

City Properties Real Estate

UAE Department of Health & Medical Services

COMSTECH

Abu Dhabi Food Control Authority

Dubai Silicon Oasis Authority

Pearson Education

PLENARY SESSION PRESENTATIONS

The Quest for Elementary Particles

John Ellis

CERN, Department of Physics, TH Division
Geneva, Switzerland

Abstract. This talk describes past progress in probing the structure of matter and the content of the Universe, which has led to the Standard Model of elementary particles, and the prospects for establishing new physics beyond the Standard Model using the LHC particle collider at CERN.

WHAT IS THE NATURE OF MATTER AND THE UNIVERSE?

Particle physics addresses some of the most basic questions about the nature of matter and the content of the Universe. Specifically, it asks the questions posed by the people portrayed in a famous painting by Gauguin: 'Where are we coming from? What are we? Where are we going?'. When I was a student, I put a copy of this painting on my office wall, to remind me why I was there. In the following, I discuss the progress we have made in answering these basic questions, and the prospects for making further progress with CERN's Large Hadron Collider (LHC), in particular [1].

INSIDE MATTER

One of the most remarkable discoveries by physicists is that all matter is made of the same limited number of constituents. The tasks of particle physicists are to understand what they are, and what are the fundamental forces between them. Specifically, earlier generations of physicists established that all the different chemical elements are composed of atoms, which contain electrons orbiting small nuclei that contain protons and neutrons. In the 1960s and 1970s, it was established that these protons and neutrons are composed of more elementary constituents called quarks. The task of CERN and other accelerator laboratories is now to study these quarks, electrons and other elementary particles, and the forces between them [2].

We currently distinguish four fundamental forces in nature: Electromagnetism, which is responsible for light and radio waves, and holds atoms together; the strong nuclear force, which holds nuclei together; the weak nuclear force, which is the origin of radioactivity (together, these two forces enable the Sun to shine); and gravity, which holds planets and stars together.

FROM COSMIC RAYS TO CERN

Many elementary particles were first discovered among the cosmic rays, energetic particles from outer space that were discovered a century ago to be striking the Earth continuously. In parallel with other accelerator laboratories, CERN was established in

CP1006, *High Energy Physics and Applications*, edited by J. Ellis, S. Nasri, and E. Malkawi

1954 to study these particles in detail, using controlled beams of identified particles with known energies. Experiments at these accelerators have made it possible to probe matter at very small distance scales, but these super-microscopes require very high energies, and hence have grown very large.

All accelerators are based on the same principles as old-fashioned television sets, before LEDs and plasma displays. They accelerate and direct particle beams using electric and magnetic fields, and then make them collide either with a fixed target, like an old-fashioned television screen, or another beam of particles traveling in the opposite direction.

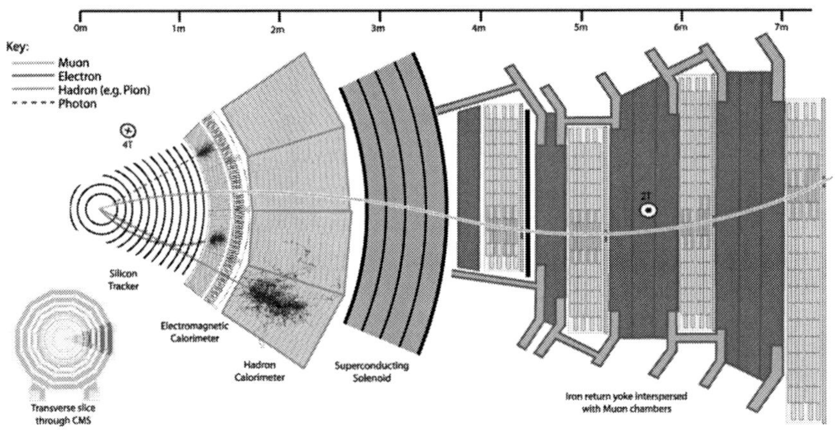

FIGURE 1. A sectional view of a typical particle detector, showing different layers of its 'onion' structure that are specialized to identify different types of particles [3].

Around the points where particles collide are constructed detectors. As seen in Fig. 1, these are shaped like onions, with many layers, each layer designed to detect a different type of particle. A modern particle detector will frequently include first components to measure the trajectories of charged particles, then calorimeters to absorb the energies of electromagnetically- and strongly-interacting particles, and finally chambers to measure the momenta of highly-penetrating particles such as muons, which are heavier partners of electrons.

LANDMARKS IN PARTICLE PHYSICS

Particle physics may be said to have started in 1897, with the discovery of the electron discovered by Thomson. This was followed in 1910 by the discovery of the nucleus discovered by Rutherford. After the proton and neutron, many other nuclear particles were discovered in the 1940s, 1950s, 1960s and 1970s. It was suggested that these could be classified easily if they were all composed of quarks, and direct experimental evidence for quarks came in the 1970s [4].

A new type of 'neutral' weak force was discovered at CERN in 1973, and this gave birth to our successful current 'Standard Model' of the elementary particles and their interactions. A key step came in 1983 with the discovery of the W and Z bosons at CERN [5]: these particles carry the weak interactions, much as the photon postulated by Einstein carries the electromagnetic interactions, and their discovery confirmed ideas for the unification of the weak and electromagnetic forces. The period between 1989 and 2000 saw the 'Standard Model' established in experiments at CERN and elsewhere, notably by the discovery of the last ('top') quark at FNAL in the US in 1995 [6]. Starting in 2008, the next steps will be taken using the LHC collider at CERN.

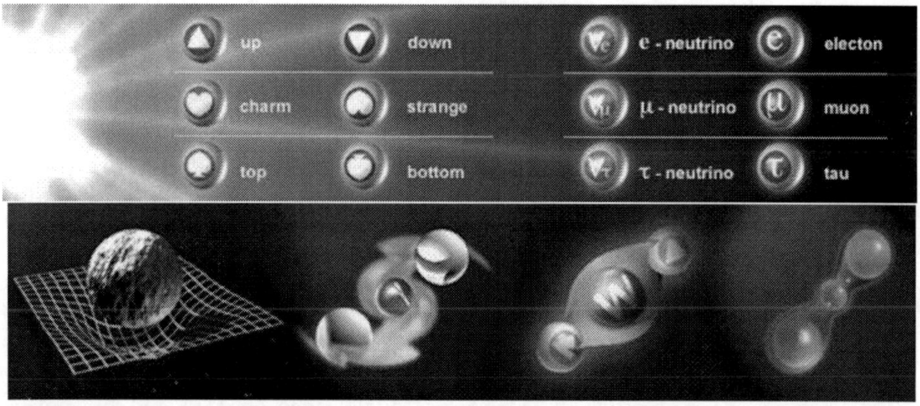

FIGURE 2. The matter particles of the Standard Model (upper panel), and the fundamental particles and their carrier particles (lower panel).

The 'Standard Model' of particle physics was proposed by Abdus Salam, Sheldon Glashow and Steven Weinberg in the 1960s, and its predictions agree with all confirmed laboratory experiments. As shown in Fig. 2, matter is composed of the following elementary particles: six types of quark, the electron and two heavier partners, the muon and tau, and three types of neutrino (emitted in weak radioactive decays). The electromagnetic forces between these elementary particles are carried by photons, the weak interactions by the W and Z bosons, the strong interactions by gluons and gravity (presumably) by the graviton, which has yet to be detected. These constituents contain all the information required to assemble all the visible matter in the Universe, and hence can be regarded as the 'cosmic DNA'.

STATUS OF THE STANDARD MODEL

As already mentioned, this Standard Model is in perfect agreement with all confirmed accelerator data, at an accuracy often measured at the per mille level. However, its success requires a missing ingredient: the consistency of the Standard Model with all the precise data on the electromagnetic and weak interactions obtained at

CERN and elsewhere requires the existence of another, as yet unseen, particle, the Higgs boson. The high degree of agreement between the data and the Standard Model seems not only to require a Higgs boson, but also to require that it be quite light [7], with a mass less than about 150 GeV, placing it well within the expected reach of the LHC.

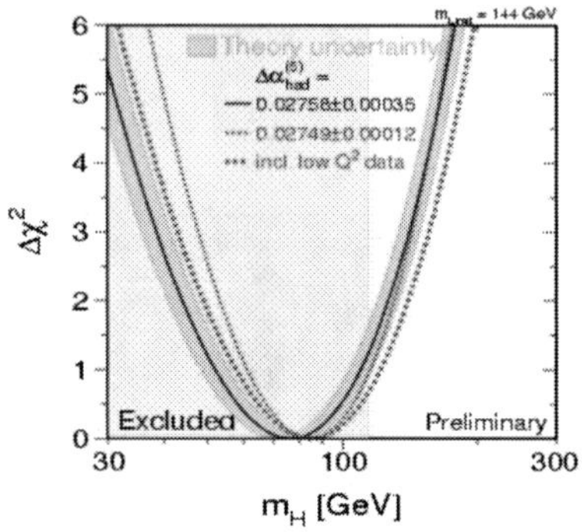

FIGURE 3. Direct searches (yellow) exclude a Higgs boson weighing below 114 GeV, and indirect evidence (blue band) favours a mass below 150 GeV [7].

OPEN QUESTIONS BEYOND THE STANDARD MODEL

However, the success of the Standard Model leaves open several fundamental questions. What is the origin of particle masses, and are they due to a Higgs boson, as postulated in the Standard Model, or are they due to some more complex mechanism? Why are there so many types of matter particles, and why do their weak interactions follow a distinctive pattern? What is the nature of the invisible dark matter that is postulated by astrophysicists, and apparently not composed of Standard Model particles? Can all the fundamental forces be unified? Is there a quantum theory of gravity? As we shall see, the LHC will address all these questions, and may well provide definite answers [8].

The most pressing question beyond the Standard Model is the origin of mass. Some particles have mass, but some do not: where do these masses come from, and why do some particles miss out? Newton showed us that weight is proportional to mass, and Einstein showed that energy is related to mass, but neither explained the origin of mass itself. According to the theory of Peter Higgs [9], which is used in the Standard Model, there is an invisible field throughout the Universe which is constant apart from a quantum fluctuation called the Higgs boson, an unseen particle that should be produced at the LHC.

There are problems with this Higgs theory that may be solved by postulating 'supersymmetry' [10]. This is a hypothetical symmetry that would unify matter particles and force particles, by relating particles that spin at different rates. The Higgs boson would have no spin, the electron, neutrinos and quarks have half a unit each, the photon, W and Z a full unit, and the graviton two units. Linking particles with different spins would help fix the particle masses, would help unify the fundamental forces, and could provide the dark matter wanted by astrophysicists and cosmologists.

As seen in Fig. 4, they tell us that the Universe has a strange recipe: normal matter can provide only a few percent of the total energy density, invisible dark matter should provide about 30%, and the rest is invisible 'dark energy', lurking in 'empty' space, that is not associated with any particles at all. In many theories, such as supersymmetry, the dark matter is composed of particles that weigh less than about a TeV, which we shall look for with the LHC.

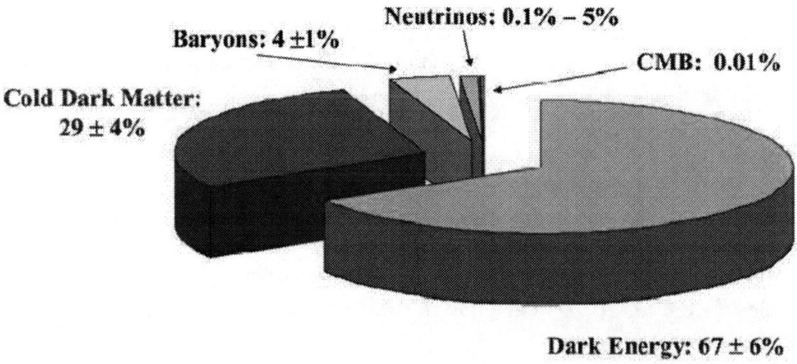

FIGURE 4. The composition of the Universe according to astrophysicists and cosmologists [11].

Another important question is: how and why do matter and antimatter differ? Dirac predicted the existence of antiparticles with the same masses as ordinary particles, but opposite internal properties such as electric charge. Antiparticles were discovered in the cosmic rays and studied using accelerators, which found that matter and antimatter are not quite equal and opposite [12]: why? Is this small difference linked to the fact that the Universe mainly contains matter, not antimatter? Experiments at the LHC and elsewhere are looking for answers to these questions.

It was Einstein's dream to unify the fundamental interactions, but he never succeeded. Many hopes for constructing this 'Theory of Everything' rest with string theory, which is discussed later in this talk.

As already mentioned, many of the answers to these fundamental questions may be accessible directly to the LHC, though some may be accessible only indirectly, and answering them may require input from astrophysics and cosmology.

The Large Hadron Collider (LHC)

The LHC is designed primarily to collide protons, each with an energy of 7 TeV, up to 1,000,000,000 times per second. Its primary scientific targets are the origin of mass, the nature of dark matter and the comparison between matter and antimatter [8]. In addition, the LHC will collide beams of heavy nuclei and probe the nature of the primordial quark-gluon plasma that filled the Universe when it was less than a microsecond old.

The LHC is located in a tunnel 27km in circumference and about 100m deep. Around the ring there are four collision points, each with a large detector. The two largest are the CMS detector [3], which is heavier than the Eiffel Tower and would sink in water, and the ATLAS detector [13], which is considerably larger, and rather lighter than the Eiffel Tower, and would float in water (though this experiment is not recommended!). CMS and ATLAS will be looking for new physics beyond the Standard Model, such as the Higgs boson and supersymmetry. Two smaller and more specialized detectors, ALICE [14] and LHCb [15], will be examining heavy-nucleus collisions and matter-antimatter asymmetries, respectively.

The event rates in ATLAS or CMS will be huge, thanks to the very high energy and collision rate of the LHC. The LHC will be a factory for many types of both old and new physics, such as the W and Z bosons and the top quark, as well as the Higgs boson and supersymmetry. CMS and ATLAS will be able to discover new particles with masses up to 5 TeV. Detailed simulations have shown that ATLAS and CMS (see Fig. 5) should be able to detect the Higgs boson in many different ways, whatever its mass [3, 13]. Simulations have also shown that they should be able to detect the classic supersymmetric signature of missing transverse energy carried away by unseen dark matter particles (see Fig. 6), which should stick out above the background due to Standard Model physics.

FIGURE 5. *A simulation how a Higgs boson might appear at the LHC [3].*

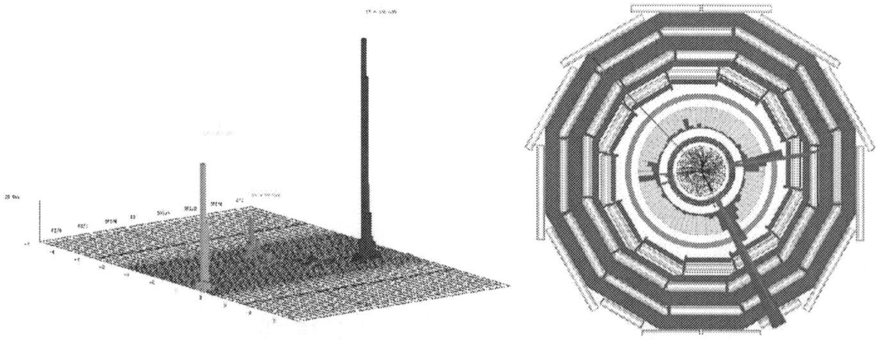

FIGURE 6. A simulation how supersymmetry might appear at the LHC [3].

SEARCH FOR A THEORY OF 'EVERYTHING'

The two greatest achievements of early 20th-century physics were quantum mechanics and Einstein's General Relativity theory of gravity. However, almost a century later, they have still not been combined. Many theorists think that string theory is the only game in town for achieving this synthesis [16]. Encouragingly, string theory seems to need both supersymmetry and extra dimensions of space, though it does not provide clear clues what energies may be required to observe them. In addition, string theory provides a powerful new theoretical toolbox that is much more powerful than traditional quantum field theory, and has many other potential applications, for example to the quark-gluon plasma [17].

How large could the extra dimensions be? It was thought originally that they would be unobservably small, but then it was suggested that they might be large enough to play roles in giving masses to supersymmetric or other particles, in which case they might be detectable at the LHC. More recently, it was suggested that there might be an extra dimension as large as a micron, or possible even infinite in size. There are many possible experimental signatures of such large extra dimensions, including the production of microscopic black holes at the LHC. These would decay very rapidly into jets of particles, electrons, neutrinos, etc., as seen in Fig. 7.

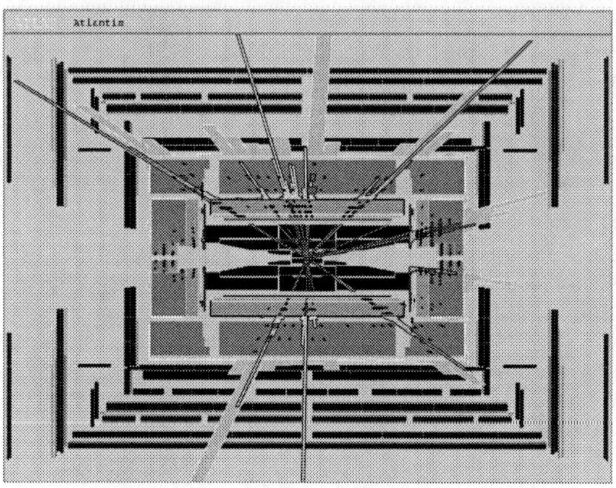

FIGURE 7. A simulation how a microscopic black hole might appear at the LHC [13].

SCIENTISTS FROM AROUND THE WORLD COLLABORATE AT CERN

Experiments at CERN are carried out mainly by teams of scientists from around the world, who remain based in their home universities and research institutes. Over 9000 scientists from dozens of non-Member States as well as CERN's 20 European Member States are registered as working at CERN. Most of these collaborations are within the frameworks of international agreements between CERN and over 40 non-Member States, including 8 in Latin America, 3 in Africa, and 13 in Asia and Australasia. Several of these countries are represented at this meeting.

It was in order to link these large international teams of researchers that the World-Wide Web was invented at CERN. The Web provides global access to documents stored on remote computers. Now CERN and its partners are deploying the World-Wide Computing Grid [18], which provides remote access to all types of computer resources, including CPUs, storage units, applications software, data, etc. It creates on demand powerful virtual computing systems that can be used to solve many problems in other areas of science and beyond.

SUMMARY

The origin of mass is the most pressing problem in particle physics today, and requires a solution at some energy accessible to the LHC. Will the answer be a simple Higgs boson or something else? Is the Higgs boson accompanied by supersymmetry? Only experiments at the LHC can provide the answers! There are plenty of speculative ideas for other physics beyond the Standard Model, such as unification, strings and extra dimension, and the LHC may also probe these speculations. We do not know what the LHC will find, but we can be sure that its discoveries will set the agenda for future projects in the global quest for elementary particles.

REFERENCES

1. J. Ellis, 'Gauguin's questions in particle physics: Where are we coming from? What are we? Where are we going?', Plenary talk at the International Europhysics Conference on High Energy Physics (EPS-HEP2007), Manchester, England, 19-25 July 2007, e-Print: arXiv:0710.5590 [hep-ph].
2. http://public.web.cern.ch/public/
3. http://cms.cern.ch/
4. F.E. Close, 'The New Cosmic Onion : Quarks and the Nature of the Universe', (CRC Press, 2006).
5. UA1 Collaboration, G. Arnsion et al., Phys. Lett. B122, 103-116, 1983; Phys.Lett.B126:398-410,1983.
6. CDF Collaboration, F. Abe et al., Phys. Rev. Lett. 74, 2626-2631, 1995.
7. http://lepewwg.web.cern.ch/LEPEWWG/
8. J. Ellis, Nature 448, 297-301, 2007.
9. P. Higgs, Phys. Lett. 12, 132-133, 1964.
10. J. Ellis, 'Outlook from SUSY07', to appear in the proceedings of the 15th International Conference on Supersymmetry and the Unification of Fundamental Interactions (SUSY07), Karlsruhe, Germany, 26 July - 1 Aug. 2007, e-Print: arXiv:0710.4959 [hep-ph].
11. N.A. Bahcall, J.P. Ostriker, S. Perlmutter and P.J. Steinhardt, Science 284, 1481-1488, 1999.
12. J.H. Christenson, J.W. Cronin, V.L. Fitch and R. Turlay, Phys. Rev. Lett. 13, 138-140, 1964.
13. http://atlas.ch/
14. http://aliceinfo.cern.ch/

15. http://lhcb.web.cern.ch/lhcb/
16. B.R. Greene, 'The elegant universe : superstrings, hidden dimensions, and the quest for the ultimate theory', (Norton, 1999).
17. S. Abreu et al., 'Heavy Ion Collisions at the LHC - Last Call for Predictions', e-Print: arXiv:0711.0974 [hep-ph].
18. http://lcg.web.cern.ch/LCG/

CERN and the Hunt for Elementary Particles and Forces

Emmanuel Tsesmelis

CERN, TS Department, CH-1211 Geneva 23 Switzerland

Abstract. CERN is the European Laboratory for Particle Physics, the world's largest particle physics research centre. Founded in 1954, the Laboratory was one of Europe's first joint ventures and has become a premier example of international collaboration. CERN's subject of study is pure science and is concentrated on exploring the Universe's most fundamental questions, such as What is it made of? and How did it come to be the way it is? The Laboratory's tools, the particle accelerators and particle detectors, are amongst the world's largest and most complex scientific instruments. The Laboratory's primary aims will be presented and a look at past achievements and present endeavours, particularly the Large Hadron Collider (LHC), will be reviewed. A brief look into the future will also be given.

Keywords: Particle physics, CERN, detector, accelerator, computing.
PACS: 10

INTRODUCTION

CERN [1] is the European Laboratory for Particle Physics, the world's largest particle physics centre and its facilities sit astride the Franco-Swiss border near Geneva, Switzerland. The Laboratory is an international centre for particle physics research where scientists from around the world come together to study the building blocks of matter and the forces that hold them together. Founded in 1954, the Laboratory is one of Europe's first joint ventures and from the outset was envisioned to be an institution where it would be possible to carry out fundamental research beyond what could be undertaken by any single national laboratory. Over the years, the Laboratory has been welcoming scientists from all continents, and at present a large component of the scientific users of CERN facilities come from its Non-Member-State partners, including scientists from the wider Middle East region, bringing with them intellectual, technical and financial contributions. Since its creation, CERN has made many important discoveries and inventions, for which CERN scientists have received prestigious awards, including Nobel prizes in physics.

CP1006, *High Energy Physics and Applications*, edited by J. Ellis, S. Nasri, and E. Malkawi

SOME MILESTONES IN CERN'S PHYSICS PROGRAMME

The charged W and neutral Z particles carry one of Nature's fundamental forces, the weak force, which causes one form of radioactivity and enables the stars to shine. By 1974, physicists at CERN had already found indirect evidence for the existence of the Z particle in the so-called `Neutral Currents' observed for the first time by a team of physicists led by André Lagarrigue working with the Gargamelle bubble chamber at CERN [2]. A few years later, the large UA1 [3] and UA2 [4] experiments at the Super Proton Synchrotron (SPS) Collider represented the accumulation of many years of knowledge and experience in the design, construction and operation of particle physics experiments. Carlo Rubbia of CERN proposed turning the Laboratory's SPS accelerator into a particle collider and by colliding counter-rotating proton and antiproton beams head on enough energy would be concentrated to produce the heavy W and Z bosons [5]. Proton-antiproton collisions at sufficient rates were made possible, in particular, through Simon van der Meer's invention of stochastic cooling to produce sufficiently dense beams of antiprotons [6]. In 1983, CERN announced the discovery of the particles known as the W and Z, a discovery that brought the laboratory its first Nobel Prize in Physics in 1984 [7]. These discoveries provided convincing evidence for the so-called electroweak theory which unifies the weak force with the electromagnetic force, and which is the cornerstone of the modern Standard Model of particles and forces.

At about the same time, a new collider for CERN was approved by the CERN Council in 1981. The Large Electron Positron Collider (LEP) [8] was to be built in a new 27 km underground tunnel. When the LEP Collider started operation in 1989 it accelerated electrons and positrons to a total energy of 45 GeV each to enable production of the Z boson, which has a mass of approximately 91 GeV. The accelerator was upgraded later to enable production of a pair of W bosons, each of approximately 80 GeV mass. The LEP Collider energy eventually topped at 209 GeV at the end in 2000. Amongst the milestones of the LEP experiments have been a) the precision measurements of the Z resonance mass, width, cross-section and branching ratios, b) the precision measurements of the W boson mass and properties and c) the determination of the number of neutrino generations. At the end of 2000, LEP was shut down and then dismantled in order to make room in the tunnel for the construction of the Large Hadron Collider (LHC).

CERN also has a long history in the development of particle detector technologies. George Charpak, a physicist at CERN since 1959, has been awarded the Nobel Prize for Physics in 1992 for his invention and development of particle detectors and in particular the multi-wire proportional chamber [9]. He has been quoted as saying that `Every step achieved in the construction of particle accelerators, such as the increase of energy and intensity, requires new detectors. Combined with the fundamental character of the questions raised by high energy physics, a formidable stimulation exists for the development of new instruments.' It is within this exceptional environment that many new detector technologies have been developed, much of which has gone on to be used in various applications beyond particle physics such as in medical imaging and precision microscopy.

THE LARGE HADRON COLLIDER

The main objectives of the LHC are the exploration of the Standard Model and the search for new phenomena beyond. In addition to the study of the Higgs sector and the origin of mass, experiments at the LHC will investigate the nature of dark matter, the primordial plasma and the asymmetry in matter and antimatter. To do this, the accelerator will provide beams of protons at unprecedented energies of 7 TeV per beam, colliding them with an equally impressive luminosity of 10^{34} cm^{-2} s^{-1}.

FIGURE 1. The CERN accelerator complex, including the 27 km circumference LHC ring.

The LHC accelerator [10], see Figure 1 and Figure 2, is built around powerful superconducting magnets which keep the LHC's proton beams in their circular orbits around the ring and which provide the focusing mechanism that keeps the particles centred on the design orbit and reduce their transverse size at the collision regions to increase the luminosity delivered to the experiments. The combination of the length of the existing tunnel and of the required beam energy results in a bending dipole field of 8.4 T. In order to minimize the number of interconnections and therefore the space lost for dipole field installations, the LHC uses large 30-t. and 15-m long dipole magnets manufactured using a niobium-titanium (NbTi)-based superconducting cable. To confine the two counter-rotating proton beams, two separate magnet apertures with opposite field orientations are produced and in order to save space a novel two-in-one magnet scheme has been adopted in which the two magnet coils have a common infrastructure and cryostat. These magnets are twice as long as dipole magnets at previous accelerators, such as the Tevatron and HERA, and the magnets are cooled by superfluid helium operated down to 1.9 K.

FIGURE 2. Photo of the LHC accelerator as installed in the ex-LEP tunnel. Some of the dipole bending magnets are shown in the foreground with their blue casings.

Four main experiments are approaching the end of their installation at the LHC. ATLAS [11] and CMS [12] are large general-purpose experiments. LHCb [13] will study b-quark systems produced primarily in the forward direction and ALICE [14] is designed specifically for studies of heavy-ion collisions. In addition, two smaller experiments – TOTEM [15] and LHCf [16] – will study particular aspects of proton–proton collisions in the forward regions of the LHC. The former will measure the total proton-proton production cross-section while the latter will make a measurement of photons and neutral pions in the very forward region of the LHC.

An important aspect of the LHC is the high-performance computing required by the LHC experiments. Due to the rarity of production of interesting events, such as the Higgs boson(s), the LHC will have a high collision rate – where up to 20 collisions are expected every 25 ns. The data transmitted in one second of LHC running is comparable to the information exchanged by the world telecommunications industry (equivalent to 100 million phone calls) resulting in about 10-15 PBytes of data per year recorded for further offline analysis. As CERN alone cannot provide the necessary resources to handle such a large data volume, computing centres which in the past have been operating in isolation will now be connected in a network thus uniting the computing resources of the world-wide community of particle physics. This scheme, for which the LHC requires the computing power equivalent to about 100,000 of today's fastest PC processors, is called the World-wide LHC Computing Grid (WLCG) [17]. Whereas the CERN-invented World Wide Web (WWW) provides seamless access to information that is stored in many millions of different geographical locations, the Grid is an emerging infrastructure that will provide seamless access to computing power and data storage capacity distributed across the globe.

THE NON-LHC PROGRAMME

The CERN physics programme extends beyond that of the LHC and includes several experiments at the Proton Synchrotron (PS) and Super Proton Synchrotron (SPS) accelerator complexes. The experiments performed at these machines are fixed-target experiments, so-called because a beam impinging on a target is used to produce events to be studied in detectors downstream. The NA48 experiment [18] at the SPS has had as its primary objective the search for direct CP violation in the K-meson sector. The discovery of direct CP violation was announced by the collaboration in 1999 and it stands as one of the most important results obtained at CERN. The next generation of studies, concentrating on very rare K-meson decays, is presently underway. The COMPASS experiment [19] at the SPS is studying hadron structure and performing hadron spectroscopy with high intensity muon and hadron beams. Experiments at the Antiproton Decelerator (AD) [20] of the PS Complex include magnetic traps to confine anti-hydrogen in order to study the properties of the simplest element in the mirror world of anti-matter. The longer-term objectives of the AD include a detailed spectroscopic study with anti-hydrogen. An extracted proton beam at the SPS is also used to create a neutrino beam sent to the Gran Sasso National Laboratory in Italy some 730 km. away (see Figure 3). As neutrinos are mixtures of three mass eigenstates, after some time the predominantly-produced ν_μ from the target at CERN may have oscillated into a ν_τ and detectors at Gran Sasso have been designed to detect the resulting τ produced in the detector target through its decay products [21]. The ISOLDE facility [22] at the PS Complex provides an online separation of isotopes used to study the properties of special radioactive nuclei while a high-intensity proton beam incident on a lead target is used to produce low-energy spallation neutrons for high-precision measurements of nuclear reactions at the Neutron Time-of-Flight (nTOF) facility [23] of the PS Complex.

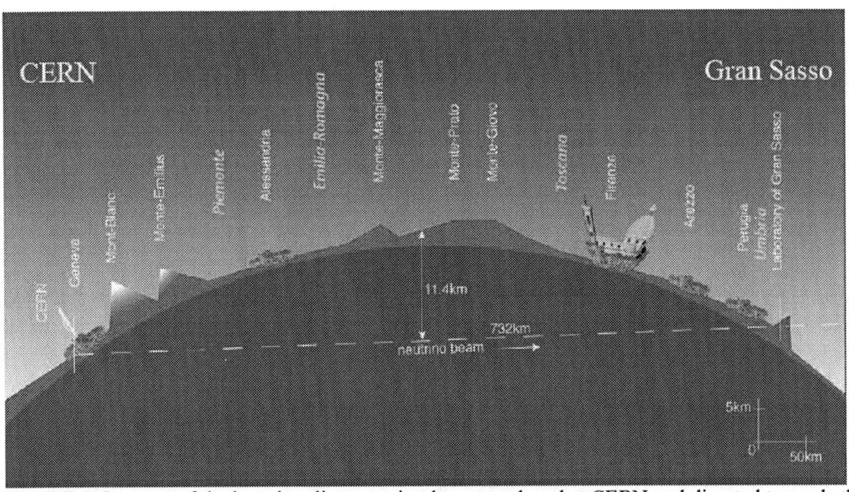

FIGURE 3. Lay-out of the long baseline neutrino beam produced at CERN and directed towards the Gran Sasso Laboratory in Italy.

POSSIBILITIES FOR THE FUTURE

Over the next several years, the LHC will provide precision tests of the Standard Model and physics beyond the Standard Model will be explored to an unprecedented reach. The main motivation to upgrade the luminosity of the LHC is to explore the physics beyond the Standard Model while at the same time completing the Standard Model physics started at the LHC. Amongst the physics issues to be addressed at an upgraded LHC are: precision Standard Model physics such as anomalous gauge boson couplings, Standard Model Higgs boson physics, supersymmetry, strong electroweak symmetry breaking, new gauge bosons, compositeness, and extra dimensions. In order to extend the reach of the LHC to high-mass systems in these areas and to make precision measurements on rare processes, an increase in the luminosity is seen as being imperative and studies are underway whereby the luminosity of the LHC will be increased up to 10^{35} cm^{-2} s^{-1} [24].

Furthermore, CERN has embarked on a feasibility study of a novel new collider called the Compact Linear Collider (CLIC) [25]. Figure 4 shows the CLIC lay-out for the particular case of a 3 TeV centre-of-mass energy. Many of the open questions from the LHC could be best addressed by such an electron-positron collider in which all the centre-of-mass energy is made available for collisions between the colliding elementary particles. A road-map has been drawn up to complete the R&D over the next years in order to demonstrate the technical feasibility of a multi-TeV centre-of-mass CLIC machine. One of the novelties of the CLIC concept resides in the mechanism used to generate the radiofrequency power that produces the accelerating field. This is done through high-intensity but low energy `drive beams' of electrons running parallel to the main beam in which the power is extracted by specially-designed decelerating structures. CLIC has unique prospects for finding new particles that do not have strong interactions and one example is the discovery of a new Z' boson which arises in models with new strong interactions or large extra dimensions.

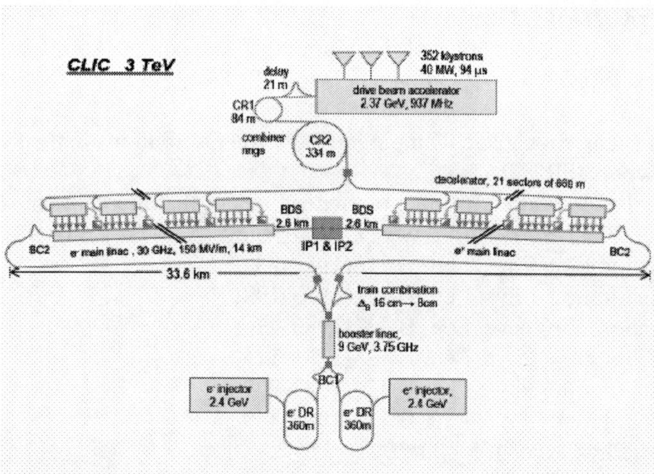

FIGURE 4. Overall lay-out of CLIC for a centre-of-mass energy of 3 TeV.

CONCLUDING REMARKS

In the world of particle physics, higher energy and intensities/luminosities as well as novel detection techniques are a key to allowing further discoveries. With the LHC, the world particle physics community has the opportunity to address fundamental questions such as What is the origin of the mass of particles? and What is the nature of dark matter? The LHC will be the most powerful instrument ever built to investigate properties of particles and the physics results from the LHC will determine the future course of high energy physics. In the meantime, future projects are being developed, including increasing the luminosity of the LHC and designing a completely new and novel accelerator – CLIC – to lead particle physics well into the next decades of fundamental research.

ACKNOWLEDGMENTS

The author would like to thank the organisers of the workshop for the invitation to make this contribution and for running smoothly a very interesting and useful workshop.

REFERENCES

1. http://www.cern.ch
2. F J Hasert *et al.,* Phys. Lett. B 46(1973)121.
   F J Hasert *et al.,* Phys. Lett. B 46(1973)138.
   F J Hasert *et al.,* Nucl. Phys. B 73(1974)1.
3. A. Astbury *et al.*, A 4π Solid-angle Detector for the SPS used as a Proton-Antiproton Collider at a Centre-of-Mass Energy of 540 GeV, Proposal, CERN-SPSC 78-6/P92, 1978.
4. B. Mansoulie, in: Proceedings of the Moriond Workshop on Antiproton-Proton Physics and W Discovery, La Plagne, Savoie, France, Editions Frontieres, 1983, p. 609.
5. S. van der Meer, Stochastic Damping of Betatron Oscillations in the ISR, CERN Internal Report ISR-PO/72-31.
6. C. Rubbia, P. McIntyre, D. Cline, in Proceedings of the International Neutrino Conference, Aachen, 1976, Viewweg, Braunschweig, 1977, p.683.
7. UA2 Collaboration, Phys. Lett. B 118(1982)203;
   UA1 Collaboration, Phys. Lett. B 122(1983)103;
   UA2 Collaboration, Phys. Lett. B 122(1983)476;
   UA1 Collaboration, Phys. Lett. B 123(1983)115;
   UA1 Collaboration, Phys. Lett. B 129(1983)273;
   UA1 Collaboration, Phys. Lett. B 134(1984)469.
8. K. Hubner, Phys. Rep. 403-404(2004)177.
9. G. Charpak *et al.*, The Use of Multiwire Proportional Chambers to Select and Localize Charged Particles, Nucl. Instr. And Meth. 62(1968)262.

10. LHC Design Report Vol. I, II, III, CERN-2004-003.
11. ATLAS Technical Design Reports and reference therein, http://atlas.web.cern.ch/Atlas/index.html
12. CMS Technical Design Reports and reference therein, http://cms.cern.ch/
13. LHCb Technical Design Reports and references therein, http://lhcb.web.cern.ch/lhcb/
14. ALICE Technical Design Reports and references therein, http://aliceinfo.cern.ch/
15. TOTEM Technical Design Report and references therein, http://totem.web.cern.ch/Totem/
16. LHCf Technical Design Report and references therein, CERN-LHCC-2006-004, LHCf-TDR-001.
17. LCG Technical Design Report and references therein, http://lcg.web.cern.ch/LCG/
18. http://na48.web.cern.ch/NA48/
19. http://wwwcompass.cern.ch/
20. http://athena.web.cern.ch/athena/
http://hussle.harvard.edu/~atrap/
http://asacusa.web.cern.ch/ASACUSA/
http://pcathena04.cern.ch/ACE
http://alpha.web.cern.ch/alpha/
21. http://proj-cngs.web.cern.ch/proj-cngs/CNGSexchangeLNGS/CNGS-LNGS.htm
http://operaweb.web.cern.ch/operaweb/index.shtml
22. http://isolde.web.cern.ch/ISOLDE/
23. http://pceet075.cern.ch/
24. S. Abdullin *et al.*, Physics Potential and Experimental Challenges of the LHC Luminosity Upgrade, Eur. Phys. J. C 39(2005)293;
O. Bruning et al., LHC Luminosity and Energy Upgrade: A Feasibility Study, CERN-LHC-PROJETC-REPORT-626, 2001.
25. The CLIC Study Team, A 3 TeV e+e- Linear Collider Based on CLIC Technology, CERN 2000-008;
http://clic-study.web.cern.ch/CLIC-Study/Welcome.html

Higgs Boson and New Physics at the LHC

Qaisar Shafi

Bartol Research Institute, Department of Physics & Astronomy, University of Delaware, Newark, DE 19716, USA

Abstract. Finding the Standard Model scalar (Higgs) boson is arguably the single most important mission of the LHC. In addition, the LHC hopefully will do its utmost to uncover direct evidence for physics beyond the standard model. In this limited amount of space, in addition to the Higgs boson, I will very briefly discuss low energy supersymmetry and warped extra dimension.

Keywords: Physics Beyond the Standard Model, Higgs, Supersymmetry, MSSM, CSSM, Extra Dimensions
PACS: 12.60.-i,98.80.-k,11.30.Pb

STANDARD MODEL

The standard model (SM) is a remarkably successful theory for the strong, weak and electromagnetic interactions at present energies. I begin by briefly summarizing its most salient features:

$$\text{Gauge Symmetry} \quad \underbrace{SU(3)}_{\text{QCD}} \times \underbrace{SU(2) \times U(1)}_{\text{EW}}$$

$$\begin{array}{cc} \downarrow & \downarrow \\ \text{confined phase} & \text{partial higgs phase} \\ \text{asymptotic freedom} & \end{array}$$

There are 15 chiral fields per family ($i = 1, 2, 3$ - color index) , such that the gauge anomalies cancel;

$$\begin{pmatrix} u_i \\ d_i \end{pmatrix}, \; u_i^c, \, d_i^c, \begin{pmatrix} \nu_e \\ e \end{pmatrix}, \; e^c \tag{1}$$

The electroweak sector is spontaneously broken by the SU(2) Higgs doublet acquiring a non-zero vacuum expectation value:

$$\langle h \rangle \neq 0 \text{ breaks } SU(2) \times U(1) \longrightarrow U(1)_{\text{em}} \tag{2}$$
$$(cf\text{: Superconductor } U(1)_{em} \longrightarrow Z_2)$$

As a consequence, charged fermions & $W^{\pm}$, Z acquire masses. There are many experimentally verified predictions of the SM :

- Weak Neutral Currents

CP1006, *High Energy Physics and Applications*, edited by J. Ellis, S. Nasri, and E. Malkawi

- Parity Violation in Atoms
- $W^{\pm}$, Z gauge bosons, Gluon Jets
- Asymptotic Freedom
- c, t, b quarks; CP violation
- Numerous other tests (Stable proton $\tau_p \geq$ few $\times 10^{33}$yrs)
- ? Higgs boson ($m_h \gtrsim 114$ GeV)

The most outstanding still to be verified prediction of the SM has to do with the existence of a neutral Higgs boson whose mass, according to LEP II , is greater than 114 GeV . It is not too much to hope that the Higgs will be the first new particle that will be found at the LHC.

What does the SM have to say about the Higgs boson mass? The mass m_h is given by

$$m_h = \sqrt{\lambda} v, \tag{3}$$

where the quartic coupling λ satisfies the two loop renormalization group equation

$$\frac{d}{dt}\lambda = \frac{\lambda}{16\pi^2}\left[12\lambda^2 - 12h_t^2 + \cdots\right]. \tag{4}$$

From arguments based on vacuum stability and perturbativity, and assuming no new physics between M_Z and M_{Planck}, one obtains the bounds

$$0.8 \lesssim \lambda \lesssim 1.1 \tag{5}$$

which yields the bounds (See Fig. 1)

$$130 \text{ GeV} \lesssim m_h \lesssim 180 \text{ GeV}. \tag{6}$$

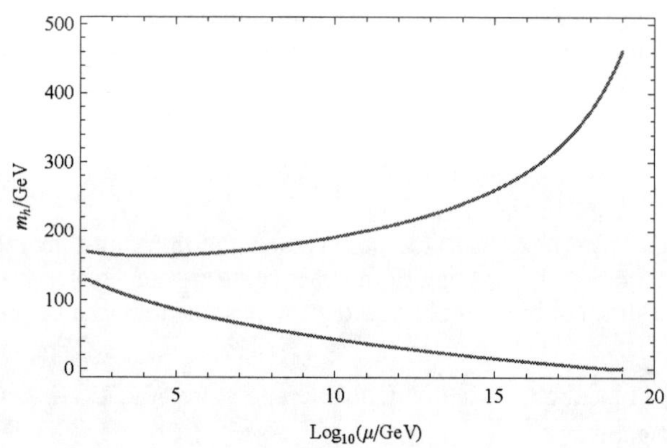

FIGURE 1. SM Higgs bounds

This bound ignores the fact that the observed neutrinos have tiny but non-zero masses, as determined from the atmospheric and solar neutrino experiments. Within the SM framework it is not possible to achieve neutrino masses , from dimension five operators, in excess of 10^{-5} eV or so. This falls well short of 0.05 eV and 0.01 eV (or so)

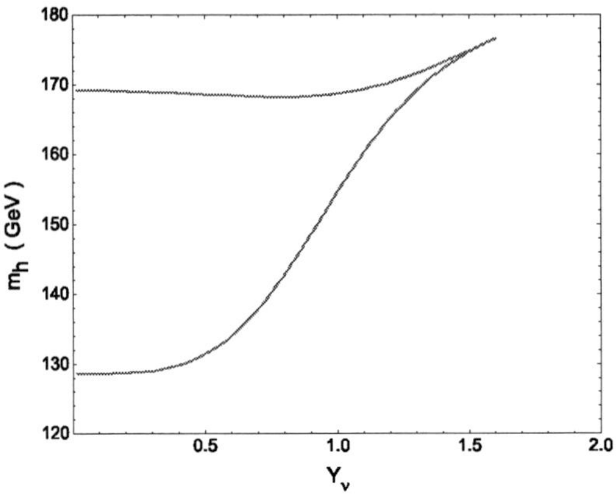

FIGURE 2. Higgs boson mass bound as a function of Y_ν. The lower and upper lines correspond to the stability and perturbativity bounds, respectively. The limit $Y_\nu = 0$ reproduces the SM result. The Higgs boson mass range is closed for $Y_\nu \simeq 1.6$.

needed to explain the observed atmospheric and solar neutrino oscillations. It is relatively straightforward to fix this problem by introducing three SM singlet (right-handed) neutrinos which help generate the desired light neutrino masses via the seesaw mechanism. I will not discuss the seesaw mechanism here but in Fig. 2 it is shown how the SM Higgs bound in Fig. 1 can be impacted by the presence of the new Yukawa coupling Y_ν which links the Higgs field, the lepton doublet and the right handed neutrino. Note that I am assuming here that only the third generation coupling plays an important role in the renormalization group analysis . For $Y_\nu = 0$ one reproduces the results of Fig. 1. Note that the Higgs mass window is closed for $Y_\nu = 1.6$ which corresponds to $m_h = 175$ GeV. For more details and additional references see a forthcoming paper by Gogoladze, Okada and Shafi [1].

SM+GENERAL RELATIVITY

Neutrino physics is by no means the sole reason for thinking that there is life (new physics) beyond the SM. The merger of the SM with Einstein's general relativity leads to the highly successful 'hot big bang cosmology' with some remarkable predictions. These include :

- Existence of blackbody radiation with a temperature of 2.725 K ;
- Redshift of galaxies;
- Nucleosynthesis.

However, the big bang cosmology fails to explain many other observations:

- Isotropy of the microwave background;

- Anisotropies in the microwave background of order 10^{-5} (or so); origin of primordial density fluctuations;
- Baryon asymmetry $(n_B/s \sim 10^{-10})$;
- Non-baryonic dark matter ($\Omega_{CDM} \approx 0.25$).

An inflationary cosmology provides the most plausible extension for overcoming some of these problems , and it turns out that new physics beyond the SM is needed to implement primordial inflation. Note that I am adopting what appears to me to be a conservative viewpoint here that modification of gravity is not required by the conundrums listed above. I have left out the fundamental issue of dark energy which is estimated by recent observations to be the dominant component today of the universe energy density. The simplest explanation is to introduce an appropriate cosmological constant term in Einstein's equations.

Before proceeding further let us list a few additional reasons for thinking about new physics beyond the SM:

- Gauge hierarchy problem (in the SM one encounters quadratic divergences in the Higgs sector so that fine tuning is needed to arrange $M_W \ll M_{Planck}$) ;
  There is no explanation in the SM for the observed hierarchical masses and mixings in the charged fermion sector and for the large mixings observed in the neutrino sector;
- Quantization of electric charge is not explained in the SM. Same holds for family replication;

SUPERSYMMETRY

MSSM

It is well known that low energy supersymmetry (SUSY) provides one of the most compelling extensions of the standard model. Here are some good reasons for this assertion based on the minimal supersymmetric standard model (MSSM) with TeV scale supersymmetry:

- Resolution (at least partial) of the gauge hierarchy problem;
- The three gauge couplings of the SM merge together at an energy scale of order 2×10^{16} GeV . This suggests SUSY grand unification which also then would explain electric charge quantization;
- The lightest supersymmetric particle due to R-parity is stable , and if it is also electrically neutral, is a highly plausible cold dark matter candidate (LSP.

One of the most attractive features of low energy SUSY is that it predicts a plethora of new particles and at least some, hopefully, will be discovered at the LHC. The MSSM features a pair of electroweak (EW) doublets and it seems reasonable to expect that the lightest CP even Higgs scalar h will be found at the LHC. Including radiative corrections, its mass is given by

$$m_h^2 \simeq M_Z^2 \cos^2 2\beta \left(1 - \frac{3}{8\pi^2}\frac{m_t^2}{v^2}t\right) + \frac{3}{4\pi^2}\frac{m_t^4}{v^2}$$

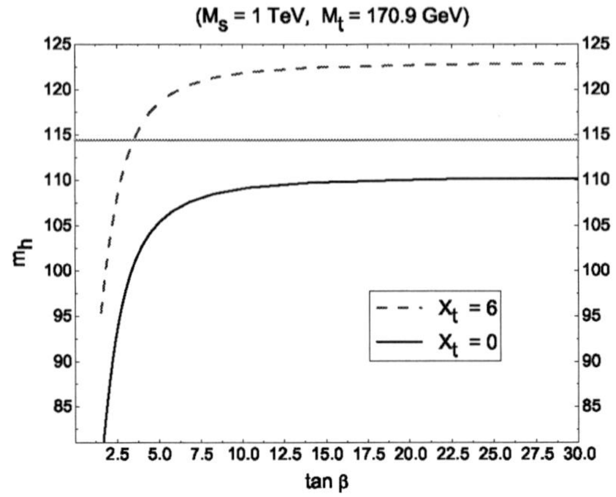

FIGURE 3. MSSM Higgs bounds

$$\times\left[\frac{1}{2}X_t+t+\frac{1}{16\pi^2}\left(\frac{3}{2}\frac{m_t^2}{v^2}-32\pi\alpha_s\right)(X_t t+t^2)\right] \tag{7}$$

where

$$t=\log\frac{M_S^2}{m_t^2},\ \widetilde{A}_t=A_t-\mu\cot\beta,\ X_t=\frac{2\widetilde{A}_t^2}{M_S^2}\left(1-\frac{\widetilde{A}_t^2}{12M_S^2}\right). \tag{8}$$

Here $M_S \sim$ TeV denotes the SUSY breaking scale and A_t is the coefficient of the soft SUSY breaking trilinear term. Fig. 3 shows the dependence of m_h on $tan\beta$, the ratio of the two EW doublet vevs. With a top quark mass of 170.9 GeV or so, it is hard to achieve a value for m_h in excess of about 125 GeV.

CMSSM

Next let us briefly discuss the so-called constrained minimal supersymmetric model (CMSSM) which has been extensively discussed in the literature. Here one makes a plausible assumption , motivated to a large extent by supergravity (SUGRA), that the soft SUSY breaking parameters of the MSSM acquire universal values at some suitably high scale, say $M_{GUT} \sim 2\times 10^{16}$ GeV. The CMSSM consists of five parameters:

$$m_0,\ m_{1/2},\ A_0,\ \tan\beta,\ \text{sign}\mu \tag{9}$$

where

m_0 = Universal soft SUSY breaking scalar mass
$m_{1/2}$ = Universal SSB gaugino mass
A_0 = Universal SSB trilinear interaction
$\tan\beta = \frac{v_u}{v_d}$
μ = Supersymmetric bilinear Higgs parameter

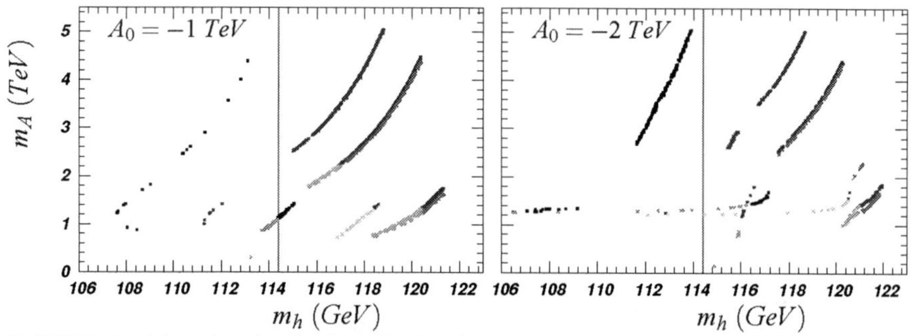

FIGURE 4. Allowed region for CP-odd Higgs boson mass m_A versus m_h. Gray, light green, light blue, and orange correspond to $\tan\beta$ =5, 10, 50 and 53 respectively and satisfy the 3σ bound on Δa_μ. Black, dark green, blue and red correspond to $\tan\beta$ =5, 10, 50 and 53, with Δa_μ outside the 3σ range.

To find mass bounds on the Higgs and sparticle masses and explore possible correlations between them, we have performed random scans using the ISAJET 7.74 package for the following range of the CMSSM parameters:

$$0 \le m_0 \le 5\,\text{TeV}, 0 \le m_{1/2} \le 2\text{TeV}$$
$$A_0 = 0.5\,\text{TeV},\ 0,\ -1\,\text{TeV},\ -2\,\text{TeV}, \tan\beta = 5,\ 10,\ 50 \text{ and } 53 \qquad (10)$$

with $\mu > 0$ and $m_t = 171.4$ GeV. In performing the analysis we have taken the following constraints into account:

- Radiative EW breaking;
- $m_h \ge 114.4$ GeV (LEP II bound);
- Lightest chargino mass > 103 GeV ;
- Ω_{CDM} consistent with WMAP observations ($\Omega_{\text{CDM}}h^2 = 0.111^{+0.011}_{-0.015}$, (2σ));
- $2.85 \times 10^{-4} \le Br(b \to s\gamma) \le 4.24 \times 10^{-4}$;
- $BF(B_s \to \mu^+\mu^-) < 1.0 \times 10^{-7}$;
- $3.4 \times 10^{-10} \le \Delta a_\mu \le 55.6 \times 10^{-10}$(where $a_\mu = \frac{(g-2)_\mu}{2}$);

Some of the results are displayed in Fig.4 and Fig.5 (for more details see [2]). A few comments are in order. A large chunk of the CMSSM parameter space is excluded by the WMAP dark matter constraints. The CP-odd scalar A is expected to have a mass in excess of about 200 GeV. Similarly, the gluino mass is estimated to be larger than about 450 GeV.

EXTRA DIMENSIONAL THEORIES

The best motivation today for higher (extra) dimensional theories comes from string theory according to which the universe may have six or even seven additional spatial dimensions which so far have escaped experimental detection. String theory provides it

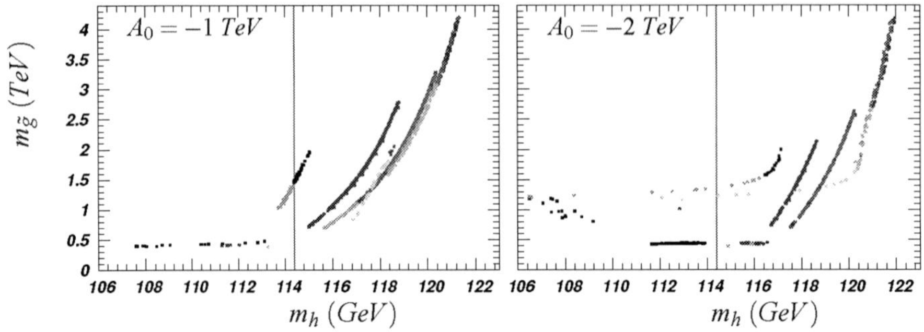

FIGURE 5. Allowed region for gluino mass versus m_h. The color code is the same as in Fig. 4.

seems the most compelling framework for unifying Einstein's gravity with the SM . If all of the extra dimensions happen to be tiny ($\sim$ Planck length) and flat, then it is most unlikely that we can 'directly' detect them experimentally.

To 'observe' one or more extra dimensions we need a great deal of cooperation from nature. Briefly, here are two options which have been extensively discussed in recent years:

Large(submillimeter size or less) Extra Dimension(s)

The scale for quantum gravity is assumed to be of order TeV, with only gravity propagating in the extra dimension(s). This bold conjecture faces serious challenges from astrophysics and cosmology but remains viable at some level. It was meant to resolve the gauge hierarchy problem but this has not been adequately demonstrated. If one or more large extra dimension is found at the LHC it would be truly remarkable and it may signal the onset of TeV scale quantum gravity. The odds for this occuring though are not terribly high.

Warped Extra Dimension

This approach attempts to resolve the gauge hierarchy problem without the introduction of a large extra dimension (See [3]). It relies on a 5D non-factorizable geometry

$$ds^2 = e^{-2\sigma(y)}\eta_{\mu\nu}dx^{\mu}dx^{\nu} + dy^2, \tag{11}$$

where $\sigma(y) = k|y|$. The 4-dimensional metric is $\eta_{\mu\nu} = \mathrm{diag}(-1,1,1,1)$, k is the AdS curvature, and y denotes the fifth dimension. This metric results from a suitable adjustment of the bulk cosmological constant and the tensions of the two 3-branes which reside at the S_1/Z_2 orbifold fixed points $y = 0, y = \pi R$. Because of the exponential ("warp") factor, the effective mass scale on the brane located at $y = \pi R$ is $M_P e^{-\pi kR}$. If $kR \sim 11$ this scale will be $\mathscr{O}$(TeV), and the brane is referred to as the 'TeV-brane'. Hence the model can generate an exponential hierarchy of scales from a small extra dimension.

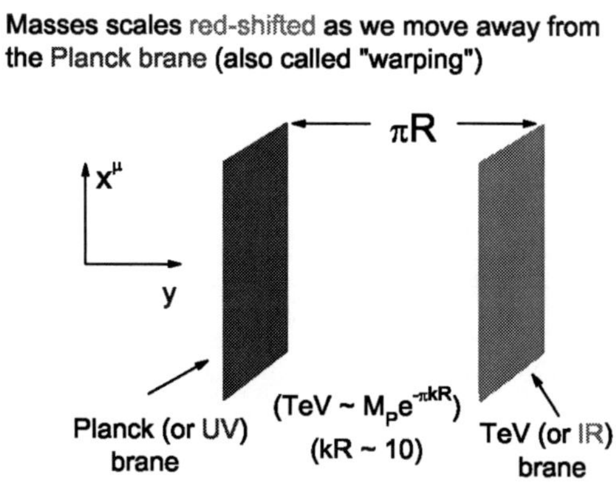

FIGURE 6. Branes in AdS Space

There are two distinct scenarios that one could consider:

- Only gravity propagates in the bulk, while the SM fields reside on the TeV brane. This approach predicts the existence of TeV scale massive gravitinos , but has to cope with several serious issues such as flavor changing neutral currents, proton stability , etc. These problems arise because of the low cutoff scale (TeV).
- In the second (more popular) approach the SM fields (except for the Higgs boson) , just like gravity, also propagate in the bulk. The most interesting consequence of this extension is the appearance of Kaluza-Klein (KK) excitations of the SM particles in the TeV mass range (See Fig. 7).

It is natural to ask whether these KK excitations can be observed at the LHC ?

This is where things start to become a bit complicated. In the simplest models with just the SM fields , the lightest KK excitations of the SM gauge bosons are expected to be in the 7-10 TeV mass range. This constraint arises by confronting the model with precision EW measurements. By considering more elaborate models which go beyond the SM framework it seems possible to achieve KK masses in the range of 2-3 TeV , and this may be more accessible at the LHC.

Let us summarize the salient features of the warped models as follows:

- Bulk SM fields;
- Higgs on the TeV Brane;
- Neutrinos could be Dirac or Majorana;
- Dirac Neutrinos: Introduce SM singlet fermion, (eliminate dim 5 Majorana masses by imposing some symmetry, say lepton number), $\Longrightarrow p$ stable; But $n-\bar{n}$ possible; Also $\mu \to e\gamma$, etc;
- Smoking Gun: KK excitations at the LHC.

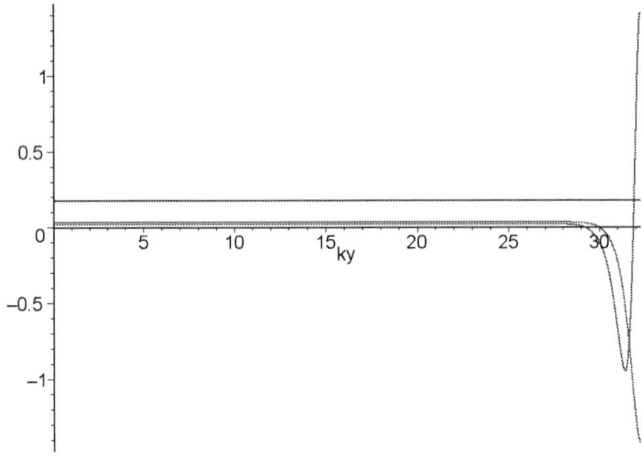

FIGURE 7. KK gauge boson wave functions, showing zero mode (blue), the first (green) and second (red) excited states.

CONCLUSION

The importance of the LHC for the future of high energy physics cannot be overemphasized. Some of the most interesting/ well motivated topics include:

- Electroweak Symmetry Breaking (Higgs, Technicolor, ...)
- Supersymmetry
- Dark Matter (LSP)
- Extra Dimensions (Kaluza Klein excitations)
- Spontaneous Parity Violation (New gauge bosons, other TeV scale particles)
- TeV Scale Quantum Gravity (Black holes,...)
- Exotic States (Magnetic Monopoles, Fractionally charged color singlets, Z flux tubes, Leptoquarks, ...).

ACKNOWLEDGMENTS

I would like to thank the organizers of the UAE-CERN workshop for arranging a very stimulating conference in a delightful setting. I also thank Ilia Gogoladze and Nobuchika Okada for an extremely enjoyable collaboration.

REFERENCES

1. I. Gogoladze, N. Okada and Q. Shafi, and references therein (in preparation).
2. A. Belyaev, S. Dar, I. Gogoladze, Q. Shafi and A. Mustafayev, [arXiv:hep-ph/0712.1049].
3. S. J. Huber and Q. Shafi, "Higgs mechanism and bulk gauge boson masses in the Randall-Sundrum Phys. Rev. D **63**, 045010 (2001) [arXiv:hep-ph/0005286].

Tev Scale Colored Particles, Baryogenesis and Dark Matter

Salah Nasri [1]

Physics Department, United Arab Emirate University,
P.O Box 17551, Al Ain, United Arab Emirates

Abstract. We propose a unified scenario of two puzzles in particle physics and cosmology, namely, the dark matter and the origin of the baryon asymmetry of the universe in a simple extension of MSSM . Our model uses a gauge singlet chiral superfield field N, where the out–of equilibrium decay of the fermionic partner of N mediated by the exchange of a colored TeV scale scalar field generates adequate baryon asymmetry for $M_N \sim 100$ GeV . The scalar partner of N is naturally the lightest SUSY particle as it has no gauge interactions and plays the role of dark matter. The model is experimentally testable in (i) neutron–antineutron oscillations with a transition time estimated to be around 10^{10} sec, (ii) discovery of colored particles X at LHC with mass of order TeV, and (iii) direct dark matter detection with a predicted cross section in the observable range in next generation dark matter search experiments.

Keywords: Particle-theory and field-theory models of the early Universe , Global symmetries, Theoretical new ideas
PACS: 98.80.Cq, 11.30.Fs, 12.90.+b

INTRODUCTION

The dark matter and the origin of matter-anti matter asymmetry in the universe are two of the major puzzles in particle physics and cosmology and require new physics beyond the standard model. It is usually assumed that these two problems are explained separately by invoking unrelated new physics. For instance, the lightest supersymmetric particle (LSP) could constitute the dark matter, while baryogenesis occurs through an unrelated mechanism involving either the decay of a heavy right–handed neutrino (leptogenesis), or new weak scale physics which makes use of the electroweak spahlerons .

The LSP dark matter in MSSM requires some tuning of parameters , either to have the right amount of Higgsino content in the LSP, or to have the right–handed stau nearly mass degenerate with the LSP to facilitate co-anninhilation .

A possible problem with leptogenesis is that in order to produce the right amount of lepton asymmetry requires that the mass of the lightest right handed neutrino to be heavier than 10^9 GeV[1] whereas the generic gravitino constraints for $m_{3/2} \leq 20$ TeV requires the reheat temperature of the Universe to be less than 10^7 GeV[2] for an unstable gravitino or 10^9 GeV for a stable gravitino. Either case causes severe tension for leptogenesis in SUSY seesaw models.

[1] email : snasri@uaeu.ac.ae

CP1006, *High Energy Physics and Applications*, edited by J. Ellis, S. Nasri, and E. Malkawi

We propose a class of models where a very minimal extension of the MSSM seems to resolve both these puzzles in a natural manner with testable consequences for the near future. Our extension of MSSM involves the addition of two new particles: an SM singlet superparticle denoted by N with mass in the 100 GeV range and an iso-singlet color triplet particle X with mass in the TeV range.

THE MODEL

As already noted, we add two new superfields to the MSSM – a standard model singlet N and a pair of color triplet with weak hypercharge $= \pm 4/3$ denoted as $X, \overline{X}$. The R–parity of the fermionic component of N is even, while for the fermionic component of X, it is odd. This allows the following new terms in the MSSM superpotential (model A):

$$W_{new} = \lambda_i N u_i^c X + \lambda'_{ij} d_i^c d_j^c \overline{X} + \frac{M_N}{2} NN + M_X X\overline{X} . \tag{1}$$

Here i, j are family indices with $\lambda'_{ij} = -\lambda'_{ji}$ and we have suppressed the color indices. An alternative possibility is to choose X to have hypercharge $-2/3$.

The fermions N and $\tilde{X}$ have masses M_N and M_X respectively. As for the scalar components of these superfields, the Lagrangian including soft SUSY breaking terms is given by

$$\begin{aligned} -\mathscr{L}_{\text{scalar}} &= |M_X|^2(|X|^2 + |\overline{X}|^2) + m_{\tilde{X}}^2|X|^2 + m_{\overline{X}}^2|\overline{X}|^2 + \\ & \quad (B_X M_X X\overline{X} + h.c) + |M_N|^2|\tilde{N}|^2 + m_{\tilde{N}}^2|\tilde{N}|^2 \\ &+ (\frac{1}{2} B_N M_N \tilde{N}\tilde{N} + h.c.) \end{aligned} \tag{2}$$

The 2×2 mass matrix in the $(X, \overline{X}^*)$ sector can be diagonalized to yield the two complex mass eigenstates X_1 and X_2 via the transformation

$$\begin{aligned} X &= \cos\theta X_1 - \sin\theta e^{-i\phi} X_2; \\ \overline{X}^* &= \sin\theta e^{i\phi} X_1 + \cos\theta X_2 \end{aligned} \tag{3}$$

where

$$\tan 2\theta = \frac{|2B_X M_X|}{|m_{\tilde{X}}^2 - m_{\overline{X}}^2|}; \ \phi = \text{Arg}(B_X M_X)\text{sgn}(m_{\tilde{X}}^2 - m_{\overline{X}}^2) . \tag{4}$$

Note that the angle θ is nearly 45^0 if the soft masses for X and $\overline{X}$ are equal. The two mass eigenvalues are

$$M_{\tilde{X}_{1,2}}^2 = |M_X|^2 + \frac{m_{\tilde{X}}^2 + m_{\overline{X}}^2}{2} \pm \left[\left(\frac{m_{\tilde{X}}^2 - m_{\overline{X}}^2}{2} \right)^2 + |B_X M_X|^2 \right]^{1/2} . \tag{5}$$

The two *real* mass eigenstates from the $\tilde{N}$ field have masses

$$M_{\tilde{N}_{1,2}}^2 = m_{\tilde{N}}^2 + |M_N|^2 \pm |B_N M_N| . \tag{6}$$

Here $\tilde{N}_1$ is the real part of $\tilde{N}$, while $\tilde{N}_2$ is the imaginary part. (A field rotation on $\tilde{N}$ has been made so that the $B_N M_N$ term is real.) With these preliminaries we can now discuss baryogenesis and dark matter in our model.

BARYOGENESIS

The matter asymmetry is generated via the post-spahleron mechanism [3]. As the universe cools to a temperature T which is below the mass of the X particle but above M_N, the X particles annihilate leaving the Universe with only SM particles and the N (fermion) and $\tilde{N}_{1,2}$ (boson) particles in thermal equilibrium. The decay of N will be responsible for baryogenesis. We therefore need to know the temperature at which the interactions of N go out of equilibrium. We first consider its decay. Being a Majorana fermion, N can decay into quarks as well as antiquarks: $N \rightarrow u_i d_j d_k$, $N \rightarrow \overline{u}_i \overline{d}_j \overline{d}_k$. The decay rate for the former is

$$\Gamma(N \rightarrow u_i d_j d_k) = \frac{6}{128} \frac{(\lambda^\dagger \lambda)\mathrm{Tr}[\lambda'^\dagger \lambda')]}{192\pi^3} \sin^2 2\theta M_N^5 \left(\frac{1}{M_{X_1}^2} - \frac{1}{M_{X_2}^2} \right)^2 . \quad (7)$$

If we take the contribution from X_1 exchange to dominate the decay, and assume that the mixing angle $\theta \simeq 45^0$. It is then easy to see that for $\sqrt{(\lambda^\dagger \lambda)\mathrm{Tr}[\lambda'^\dagger \lambda')]} \sim 10^{-3}$, N decay goes out of equilibrium below its mass. Other processes involving N such as $q + N \rightarrow \bar{q} + \bar{q}$ also go out of equilibrium at this temperature.

In order to calculate the the baryon asymmetry of the universe, we look for the imaginary part from the interference between the tree–level decay diagram and the one–loop correction arising from $W^\pm$ exchange. These corrections have a GIM–type suppression, since the $W^\pm$ only couple to the left–handed quark fields while the tree–level decay of N is to right–handed quarks. Following Ref. [3], we find the dominant contribution to be

$$\frac{\varepsilon_B}{\mathrm{Br}} \simeq \left\{ -\frac{\alpha_2}{4} \right\} \frac{\mathrm{Im}[(\lambda^* \hat{M}_u)^T V \hat{M}_d \lambda'].[\lambda'^* \hat{M}_d V^T \lambda \hat{M}_u]}{M_W^2 M_N^2 (\lambda^\dagger \lambda)\mathrm{Tr}(\lambda'^\dagger \lambda')} \quad (8)$$

where Br stands for the branching ratio into quarks plus anitquarks, and $\hat{M}_u = (m_u, m_c, m_t)^T$, $\hat{M}_d = diag.\{m_d, m_s, m_b\}$ The interesting point is that as in Ref. [3] the asymmetry is completely determined by the electroweak corrections. A typical leading term in Eq. (8) is of the form $(-\alpha_2/4)(m_c m_t m_s m_b)/(m_W^2 m_N^2)$ which yields $\varepsilon_B \simeq 3 \times 10^{-8}$ with only mild dependence on the couplings λ_i, λ'_{ij}. This can easily lead to desired value for the baryon asymmetry.

DARK MATTER

In minimal supergravity model (mSUGRA) type models, generally, one chooses a common scalar mass for all particles at the SUSY breaking scale (say M_P or M_{GUT}), so that scalar masses at the weak scale are determined by the renormalization group running.

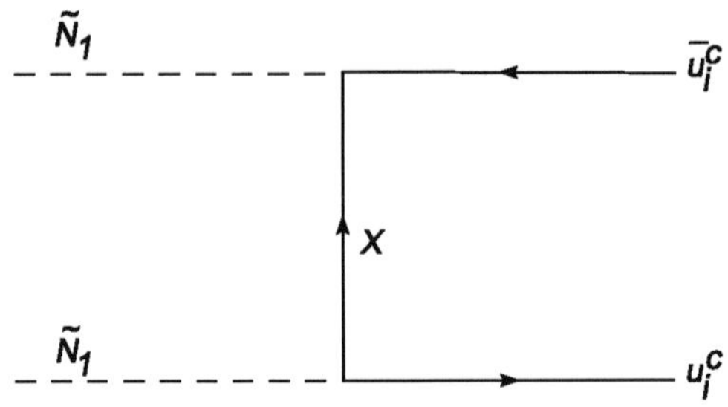

FIGURE 1. Annihilation of $\widetilde{N}$ into fermions.

There are two kinds of contributions to the running of the soft SUSY breaking masses – gauge contributions which increase masses as we move lower in scale, and Yukawa coupling contributions which tend to lower the masses as we move lower in scale. As far as the scalar $\tilde{N}_1$ particle goes since it has no gauge couplings, its mass naturally goes somewhat lower as we move from the Planck scale to the weak scale and becomes naturally the lowest stable state. Furthermore since its couplings $\lambda_{1,2}$ are in the range of 0.1-0.001, they are not strong enough to drive $m^2_{\tilde{N}_1}$ negative like the $m^2_{H_u}$.

In the early universe, the LSP $\tilde{N}_1$ will annihilate into quark-antiquark via the exchange of $\tilde{X}$ fermion. The annihilation cross section is given by

$$\sigma(\tilde{N}_1\tilde{N}_1 \to q\bar{q})v_{\rm rel} = \frac{C'(\lambda^\dagger\lambda)^2}{8\pi s}\left(\frac{a}{b}\tanh^{-1}(\frac{b}{a}) - 1\right) \tag{9}$$

where

$$a = 2E^2 - M^2_{\tilde{N}_1} + M^2_{\tilde{X}}; \quad b = 2E|\overrightarrow{p}|\,. \tag{10}$$

Here $C' = 3$ is a color factor, E and $\overrightarrow{p}$ are the energy and momentum of one of the $\tilde{N}_1$, s is the total CM energy. For $M_X \gg E$, the cross section reduces to

$$\sigma v_{\rm rel} \simeq \frac{1}{8\pi}(\lambda^\dagger\lambda)^2\frac{|\overrightarrow{p}|^2}{M_X^4} \tag{11}$$

For the coupling $(\lambda^\dagger\lambda) = 0.25$, $|\vec{p}| = 30$ GeV and $M_X = 200$ GeV, the cross section is of the order of a pb as would be required to generate the right amount of relic density.

Note that in our model, there is no need for co-annihilation but if necessary, the mass of $\tilde{N}_2$ is naturally close to $\tilde{N}_1$ if $|BM_N|$ is chosen to be small.

The scattering of $\tilde{N}_1$ (with momentum p) off a quark (with momentum k) occurs via the s-channel exchange of the fermionic component of X (see figure 2) . The amplitude is given by

$$\mathscr{M}_{\tilde{N}_1+q} = i\frac{\lambda_1^2}{4M_X^2}\overline{u(k')}\gamma^\mu u(k)Q_\mu \tag{12}$$

where, $Q = k + p$. At the nucleon level , the time component of the vector current dominates (spin-independent) over the spatial component (velocity dependent). The

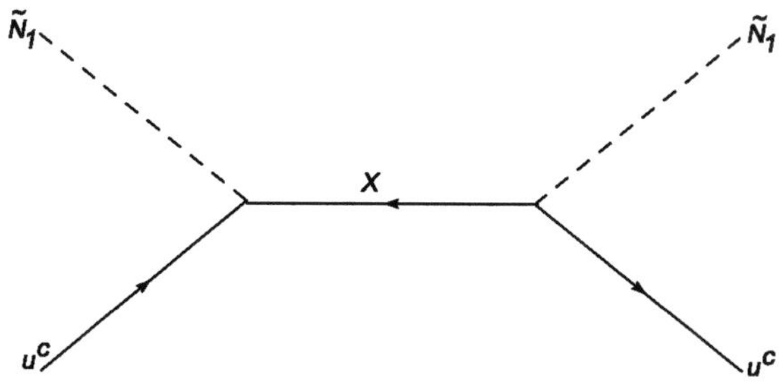

FIGURE 2. The scattering of $\widetilde{N}$ off a quark in the nucleon.

nucleon–$\tilde{N}_1$ cross section is given by

$$\sigma_{\tilde{N}_1+p} \simeq \frac{\lambda_1^4 m_p^2}{64\pi M_X^4}\left(\frac{A+Z}{A}\right)^2 \tag{13}$$

where A are the atomic numbers and charge of the nucleus and λ_1 is the coupling of N to the first generation quarks. The sum $\lambda_1^2+\lambda_2^2$ is constrained by LSP annihilation requirement but individually λ_1 is not. If we choose $\lambda_1 \sim 0.1-0.01$, then the cross section is around $10^{-43}\ cm^2 - 10^{-47}\ cm^2$ which is the range being currently explored [4].

COMMENTS

(i) One of the interesting predictions of our model is the existence of neutron-anti-neutron oscillation at an observable rate. The strength for this process (taking into account the anti-symmetry of λ' couplings) is given by:

$$G_{\Delta B=2} \simeq \frac{(\lambda_1\lambda'_{12})^2}{M_N M_X^4} \tag{14}$$

The $\Delta B = 2$ operator in this case has the form $u^c d^c s^c u^c d^c s^c$. If we choose the strange quark component in the nucleon to be about 1%, then choosing $\lambda_1\lambda'_{12} \simeq 10^{-4.5}$, we find that $G_{\Delta B=2} \simeq 10^{-27}$ GeV^{-5} which corresponds to the present limit on $\tau_{N-\bar{N}} \sim 10^8$ sec. [5, 6]. It is interesting that the expectation for the $N-\overline{N}$ transition is in the range accessible to experiments and this can therefore be used to test the model.

(ii) The X particle in our model can be searched for at the LHC. Once produced, X will decay into two jets, e.g., a b jet and a light quark jet. We point out that there is an interesting difference in discovery of SUSY at LHC in our model. Consider up type squarks pair produced at LHC. The squark will decay into a quark plus a neutralino. In our model, the neutralino is unstable, it decays into $u^c d^c d^c \tilde{N}$. So one SUSY signal will be six jets plus missing energy. The scalar up squark can also decay directly into $\tilde{N} d^c d^c$. In this case the signature will be 4 jets plus missing energy.

(*iii*) We also point out that this model is different in many respects from some other suggestions of right handed sneutrino dark matter in literature [7, 8, 9].

(*iv*) The models presented are compatible with gauge coupling unification, provided that the X particle is accompanied by other vector–like states which would make complete $10+\overline{10}$ representations of $SU(5)$. These extra particles will have no effect on baryogenesis and dark matter phenomenology.

CONCLUSION

In conclusion, we have presented a unified particle physics description of two major problems of cosmology – the baryon asymmetry of the universe and the dark matter. The desired properties of the dark seem to come out in a more natural manner than in the MSSM. A prediction of the model is the existence of neutron-anti-neutron oscillation and a new particles beyond MSSM coupling to quarks. These new particles could therefore be observable at LHC. The cross section for direct dark matter detection is predicted to be in the observable range.

REFERENCES

1. S. Davidson and A. Ibarra, Phys. Lett. B **535**, 25 (2002).
2. K. Kohri, T. Moroi and A. Yotsuyanagi, Phys. Rev. D **73**, 123511 (2006).
3. K. S. Babu, R. N. Mohapatra and S. Nasri, Phys. Rev. Lett. **97**, 131301 (2006)
4. P. L. Brink *et al.* [CDMS-II Collaboration], *In the Proceedings of 22nd Texas Symposium on Relativistic Astrophysics at Stanford University, Stanford, California, 13-17 Dec 2004, pp 2529* [arXiv:astro-ph/0503583].
5. M. Baldo-Ceolin *et al.*, Z. Phys. C **63**, 409 (1994).
6. M. Takita *et al.* [KAMIOKANDE Collaboration], Phys. Rev. D **34**, 902 (1986); J. Chung *et al.*, Phys. Rev. D **66**, 032004 (2002).
7. N. Arkani-Hamed, L. J. Hall, H. Murayama, D. R. Smith and N. Weiner, Phys. Rev. D **64**, 115011 (2001).
8. H. S. Lee, K. Matchev and S. Nasri, Phys. Rev. D **76**, 041302 (2007).
9. T. Asaka, K. Ishiwata and T. Moroi, Phys. Rev. D **73**, 051301 (2006); S. Gopalakrishna, A. de Gouvea and W. Porod, JCAP **0605**, 005 (2006); J. Mcdonald, hep-ph/0609126; S. Abel and V. Page, JHEP **05**, 024 (2006).

Supersymmetry At LHC

Shaaban Khalil

Centre for Theoretical Physics, The British University in Egypt, El Sherouk City, Postal No. 11837, P.O. Box 43, Egypt.
Department of Mathematics, Ain Shams University, Faculty of Science, Cairo, 11566, Egypt.

Abstract. One of the main motivation of the experiments at the Large Hadron Collider (LHC), scheduled to start around 2006, is to search for supersymmetric particles. The region of the parameter space of the minimal supersymmetric standard model, where supersymmetry can be discovered is investigated. We show that if supersymmetry exists at electroweak scale, it would be easy to find signals for it at the LHC. If the LHC does find supersymmetry, this would be one of the greatest achievements in the history of theoretical physics.

Keywords: supersymmetry, LHC
PACS: 11.40

INTRODUCTION

The Standard Model of elementary particle physics (SM), based on the gauge symmetry $SU(3)_C \times SU(2)_L \times U(1)_Y$, is in an excellent agreement with present experimental results. In fact, the stunning experimental confirmations of the SM that have kept accumulating throughout the last years, make it mandatory for any new physics to exactly reproduce the SM at the Fermi scale ($\sim$ 100 GeV) [1]. However, the SM cannot be the ultimate theory, valid to high energy scales (up to the Planck scale, $M_{Planck} \sim 1.2 \times 10^{19}$ GeV, where gravity is expected to become strong). In fact, there are still many important questions that the SM cannot answer. For instance, the SM cannot explain why a particle has a certain mass. Also within the framework of the SM we do not know why there are six kinds of quarks and leptons, if the electroweak and the strong forces can be unified and how gravity fits into all of this, if neutrinos have masses, why matter dominates over antimatter, and what is the dark matter made of.

Supersymmetric theories now stand as the most promising candidate for a unified theory beyond the SM. Supersymmetry (SUSY) is a generalization of the space–time symmetries of the quantum field theory that links the matter particles (the quarks and leptons) with the force–carrying particles, and implies that there are additional 'superparticles' necessary to complete the symmetry. Despite the absence of experimental verification, relevant theoretical arguments can be given in favour of supersymmetry:

(i) SUSY is a new symmetry which relates bosons and fermions. We know from the past that symmetries are crucial in particle physics and besides, supersymmetry implies a new kind of unification between particles of different spin. In this respect, the Higgs is no longer a mysterious particle as it stands in the SM, where it is introduced just in order to break the electroweak symmetry. The supersymmetric standard model is naturally full of scalars (squarks, sleptons and Higgs) related through supersymmetry with their fermionic partners (quarks, leptons and Higgsinos);

CP1006, *High Energy Physics and Applications*, edited by J. Ellis, S. Nasri, and E. Malkawi

(ii) SUSY ensures the stability of hierarchy between the week and the Plank scales. Since any fundamental theory must include gravity, due to quantum corrections it turns out that the mass of the scalar particles are proportional to the natural scale of the theory M_{Planck}. However, since these mass terms contribute to the Higgs potential, they should be of order the electroweak scale (100 – 1000) GeV, in order to avoid any type of fine-tuning when breaking the electroweak symmetry. Since supersymmetry requires that the scalar masses and the masses of their superpartner to be related, the dangerous contributions of SM particles to quantum corrections are cancelled with new ones which are present owing to the existence of additional superpartners [2]; Therefore, this problem of stabilizing the scalar masses against quantum corrections is solved in supersymmetric theories.

(iii) The local version of supersymmetry leads to a partial unification of the SM with gravity, the so-called 'supergravity', which is the low energy limit of superstrings;

(iv) While in the SM the quadratic and quartic terms of the Higgs potential, necessary to break the electroweak symmetry, have to be postulated 'adhoc', they appear in a natural way in the context of the supersymmetric standard model;

(v) If the idea of the Grand Unification of all interactions is correct, then all the interactions we observe have to merge into one force. In the SM, it was realized that the strength of these interactions approached one another at short distances or high energies. However, the joining of the three gauge couplings constant of the SM forces at a single unification scale, in agreement with experimental results from the LEP accelerator at CERN Laboratory (Geneva, Switzerland), can only be obtained assuming supersymmetry.

The price one pays for such a beautiful theory is the abundance of new particles that must be included into the theory, since, in order to satisfy the requirements of supersymmetry, each particle must have a superpartner (a boson for a fermion and vice versa). Recall that, ordinary particles and their superpartners only differ in spin, supersymmetry would imply that they have the same masses, and therefore there would be a superpartner for the electron (selectron) with the electron mass (0.5 MeV). However, there is no experimental evidence for such particles with these masses. Thus, if supesymmetry is meant to play a realistic role in particle physics, it must be broken so that the superpartners are heavier than the known elementary particles. Although our theoretical ideas are still insufficient to predict the superpartner masses, there are several arguments that lead to the conclusion that a typical superpartner mass should be in the range of 100 GeV to 1000 GeV, *i.e.*, 100 – 1000 times heavier than a proton. This is a very interesting mass range since it is just what the LHC is going to explore. Thus, the goal of the LHC will be either to discover weak–scale supersymmetry or to exclude it. If the LHC discovers SUSY, this will certainly be one of the greatest achievements in the history of theoretical physics.

THE SUPERSYMMETRIC EXTENSION OF THE STANDARD MODEL

The minimal supersymmetric standard model (MSSM) is essentially a straightforward supersymmetrization of the SM with a minimal number of new parameters. It is the most

TABLE 1. The MSSM particle content.

Particle	Symbol	Spin	Superpartner	Symbol	Spin
Quark	q	1/2	Squark	$\tilde{q}$	0
Neutrino	ν	1/2	Sneutrino	$\tilde{\nu}$	0
Electron	e	1/2	Selectron	$\tilde{e}$	0
Muon	μ	1/2	Smuon	$\tilde{\mu}$	0
Tau	τ	1/2	Stau	$\tilde{\tau}$	0
W	W	1	Wino	$\tilde{W}$	1/2
Z	Z	1	Zino	$\tilde{Z}$	1/2
Photon	γ	1	Photino	$\tilde{\gamma}$	1/2
Gluon	g	1	Gluino	$\tilde{g}$	1/2
Higgs	$h,A,H^0,H^\pm$	0	Higgsinos	$\tilde{h},\tilde{A},\tilde{H}^0,\tilde{H}^\pm$	1/2

widely studied potentially realistic SUSY model. To construct supersymmetric models one would like to have a formalism in which supersymmetry is manifest. At the beginning, the way of constructing an action which is invariant under SUSY transformation was one of trial and error. One starts with the first (lowest) component, the scalar field ϕ, in such a way that the SUSY transformation leads to $\phi \rightarrow \psi$. However, one can not close the SUSY algebra with just ϕ and ψ, hence one alters the transformation law for ψ by introducing an additional bosonic field F. When superfields were introduced by Salam and Strathdee, a systematic way of constructing SUSY invariant actions became available. The superfield is a kind of a multiplet that contains both bosons and fermions. The MSSM contains three generations of quark and lepton superfields, the superfields necessary to gauge the $SU(3)_C \times SU(2)_L \times U(1)_Y$ gauge of the SM, and two $SU(2)$ doublet Higgs superfields. The introduction of a second Higgs doublet is necessary to give masses to both up and down type quarks [4]. It is worth mentioning that 'minimal' in MSSM means that we keep the number of the superfields as small as possible. The particle content of the MSSM together with their quatnum numbers is presented in Table 1.

As a result of Baryon–Lepton ($B-L$) invariance, the MSSM possesses a discrete R-parity invariance, where $R=(-1)^{3(B-L)+2S}$ for a particle of spin S There are two remarkable phenomenological implications of the presence of R parity: (i) SUSY particles are produced or destroyed only in pairs and (ii) The lightest SUSY particle (LSP) is absolutely stable and, hence, it might constitute a possible candidate for dark matter [10]. Since the neutral LSP interacts weakly, a major signature for R-parity conserving models is represented by missing energy events (for instance, $e^+e^- \rightarrow$ jet + missing energy).

Supersymmetry cannot be an exact symmetry of nature since if it were, it would imply the existence of mass degenerate fermion boson pairs, for which there is no experimental evidence. Thus, supersymmetry must be a broken symmetry. One may introduce terms in the Lagrangian which explicitly break SUSY, but one should be careful that these SUSY breaking terms will not spoil the SUSY solution to the hierarchy problem, we discussed in the previous section. This restricted type of terms exists, and are called soft SUSY breaking terms. They have been catalogued by Girardello and Grisaru [6] as the following: masses for the scalars, masses for the gauginos, and cubic couplings for the scalars. The soft terms are very important since they determine the SUSY spectrum and contribute to the Higgs potential generating the radiative breakdown of the electroweak symmetry.

There are several MSSM models that are based on different mechanisms of supersymmetry breaking, and each predicts different experimental signatures. The model most often used to interpret experimental data is the minimal supergravity model (mSUGRA). In this model, certain universality of soft SUSY breaking terms is assumed at grand unification (GUT) scale $\sim 10^{16}$ GeV and the renormalization group equations (RGE) are used to calculate the parameters at the electroweak. Supergravity models provide a mechanism for transmitting the SUSY breaking of the hidden sector to the particle spectrum of the MSSM. In this case of gravity mediated supersymmetry breaking, gravity is the messenger of SUSY breaking and the gravitino mass is of the order of the electroweak scale. Nevertheless, and since its couplings are gravitional in strength, it plays no role at colliders.

The mSUGRA is a natural framework for studying production and decay of SUSY particles since it has only five free parameters. Thus one one can systematically study the whole parameter space of this model. A strong constraint on the parameters is given by the requirement of spontaneous electroweak symmetry breaking. Therefore, the parameters which remain are: m_0, the common scalar mass at GUT scale, $m_{1/2}$ the unifying gaugino mass, A_0 the common trilinear term, $\tan\beta = v_2/v_1$, with $v_1(v_2)$ being the Higgs vacuum expectation value of the Higgs $H_1(H_2)$, and sign of the Higgs bilinear term, $\mu H_1 H_2$ [4].

The masses of the scalar partners of quarks and leptons receive different contributions. There is the universal supersymmetry breaking term m_0^2 at GUT scale. Running from high to low energies, this parameter is normalized and assumed distinct values for different scalar partners. This is generally the most important contribution. A second contribution comes from the gauge piece of the scalar potential once the doublets $H_{1,2}$ get a vacuum expectation value. In addition, there is another contribution, which is equal to the squared mass of the respective fermionic partners, and it is essentially negligible except for the third generation squarks.

The mass spectrum for the $\widetilde{W}$,$\widetilde{B}$ and Higgsinos is more complicated because once $SU(2)_L \times U(1)_Y$ is broken they mix among themselves. This is due to the existence of the SUSY gauge couplings which give rise to mass terms that combine the charged winos with the charged Higgsinos and $\widetilde{W}^0$ and $\widetilde{B}^0$ with the neutral Higgsinos. As a result, the physical mass eigenstates are linear combinations of these states, called charginos ($\widetilde{\chi}_i^{\pm}, i = 1,2$) and neutralinos ($\widetilde{\chi}_i^0, i = 1,2,3,4$) which are obtained by diagonalizing the corresponding mass matrices that depend on $m_{1/2}$, μ and $\tan\beta$.

As it can be seen, supersymmetry introduces many new particles which can affect what we observe in the accelerators, either directly by finding a new particle or indirectly by changing some measured width or branching ratio. The main constraints on the SUSY masses that we have mentioned in the previous section comes from high energy e^+e^- or hadron collider experiments. In Table 2 we list the relevant high energy collider projects.

Today's negative SUSY searches [8] provide the following approximate lower mass limit. The most effective limit on the chargino mass comes from LEPII, through the search for the process $e^+e^- \to \widetilde{\chi}^+\widetilde{\chi}^-$ The present bound on the chargino mass from

LEPII is $m_{\tilde{\chi}^{\pm}} > 90$GeV. The neutralino can also be produced at LEP and would contribute to the invisible width of the Z boson. The limit on the lightest neutralino is given by $m^{0}_{\tilde{\chi}} \geq 23$ GeV. Strong interacting particles (squarks and gluinos) could be produced in hadronic collisions. Assuming that mSUGRA GUT assumptions hold, one can derive a limit on the squark and gluino mass [8]

$$m_{\tilde{q}} > 176\text{GeV} \qquad\qquad m_{\tilde{g}} > 160 \text{ GeV}.$$

The investigation of the low energy effects, which might arise from supersymmetry, is another possible way of testing SUSY theories. For instance, severe constraints on the MSSM, which arise upon consideration of rare processes and precision experiments. In particular, the most stringent constraints arise from the $b \to s\gamma$ decay width as measured by the CLEO experiment [9]. Further, if the dark matter in the universe is made of the lightest neutralino, then $\tilde{\chi}^0$ can be probed by different direct and indirect dark matter searches [10]. Furthermore, there are indirect evidences for SUSY through departure from the SM prediction in FCNC and CP violation processes. Indeed, these processes have potentialities to exhibit manifestations of SUSY, see for instance Ref.[11].

SUSY PRODUCTION AT LHC

It is believed that the LHC at CERN is the machine that will take particle physics into a new phase of discovery. The LHC is a proton–proton collider which will explore energies around 1 TeV. At present, six experiments are proposed to run at the LHC. The two largest, Compact Muon Solenoid (CMS) and A Toroidal LHC Apparatus (ATLAS), are general–purpose experiments that adopt different approaches. The ATLAS and CMS collaborations at LHC selected five points in the mSUGRA model for their analysis. In Table 3 we list these five points, and the corresponding masses of the SUSY particles for these five points are given in Table 4.

TABLE 2. Parameters for the five LHC SUGRA points.

Point	m_0 (GeV)	$m_{1/2}$ (GeV)	A_0 (GeV)	$\tan\beta$	sgn μ
1	400	400	0	2.0	+
2	400	400	0	10.0	+
3	200	100	0	2.0	−
4	800	200	0	10.0	+
5	100	300	300	2.1	+

At LHC the total SUSY particle production cross section are largely dominated by strongly interacting sparticles. Thus, a typical high mass SUSY signal has squarks and gluinos which decay through a number of steps to quarks, gluons, charginos, neutralinos, W, Z , Higgses and finally to a stable $\tilde{\chi}^0_1$. The two largest branching ratios are

$$B(\tilde{g} \to \tilde{\chi}_1^- t\bar{b} + h.c.) \approx B(\tilde{g} \to \tilde{\chi}_2^- t\bar{b} + h.c.) \approx 23\%$$

TABLE 3. Masses of the superpartners, in GeV, at the five LHC SUGRA points [7].

Point	1	2	3	4	5
$\widetilde{g}$	1004	1009	298	582	767
$\widetilde{\chi}_1^{\pm}$	325	321	96	147	232
$\widetilde{\chi}_2^{\pm}$	764	537	272	315	518
$\widetilde{\chi}_1^0$	168	168	45	80	122
$\widetilde{\chi}_2^0$	326	321	97	148	233
$\widetilde{\chi}_3^0$	750	519	257	290	497
$\widetilde{\chi}_4^0$	766	538	273	315	521
$\widetilde{u}_L$	957	963	317	918	687
$\widetilde{u}_R$	925	933	313	910	664
$\widetilde{d}_L$	959	966	323	921	690
$\widetilde{d}_R$	921	930	314	910	662
$\widetilde{t}_1$	643	710	264	594	489
$\widetilde{t}_2$	924	933	329	805	717
$\widetilde{b}_1$	854	871	278	774	633
$\widetilde{b}_2$	922	930	314	903	663
$\widetilde{e}_L$	490	491	216	814	239
$\widetilde{e}_R$	430	431	207	805	157
$\widetilde{\nu}_e$	486	485	207	810	230
$\widetilde{\tau}_1$	430	425	206	797	157
$\widetilde{\tau}_2$	490	491	216	811	239
$\widetilde{\nu}_\tau$	486	483	207	806	230
h^0	111	125	68	117	104
H^0	1046	737	379	858	638
A^0	1044	737	371	859	634
$H^{\pm}$	1046	741	378	862	638

However, decays into both chargino and all four neutralinos with all allowed combinations of quarks are significant, which leads to many complex signatures.

Charginos and neutralinos

As mentioned above, the mixing of the fermionic partners of the electroweak gauge and Higgs bosons, the gauginos and Higgsinos, gives rise to the physical mass eigenstates called the charginos and neutralinos. They are among the lightest SUSY particles and therefore present particular interest [5]. The two lightest neutralinos and the lightest chargino ($\widetilde{\chi}_1^0$, $\widetilde{\chi}_2^0$, $\widetilde{\chi}_1^{\pm}$) have as their largest mixing component the gauginos, and their masses are determined by the common gaugino mass, $m_{1/2}$. Within the mSUGRA scenario we have

$$\begin{aligned} M_{\widetilde{\chi}_1^0} &\approx 0.45 m_{1/2}, \\ M_{\widetilde{\chi}_2^0} &\approx M_{\widetilde{\chi}_1^{\pm}} \approx 2M_{\widetilde{\chi}_1^0} \\ M_{\widetilde{\chi}_2^0} &\approx (0.25-0.35) M_{\widetilde{g}}. \end{aligned} \tag{1}$$

The lightest chargino ($\widetilde{\chi}_1^{\pm}$) has several leptonic decay modes, producing an isolated lepton and missing energy, due to the undetectable neutrino and the LSP:

$$\begin{aligned}
\widetilde{\chi}_1^{\pm} &\rightarrow \widetilde{\chi}_1^0 l^{\pm}\nu, \\
\widetilde{\chi}_1^{\pm} &\rightarrow \tilde{l}_L^{\pm}\nu \rightarrow \widetilde{\chi}_1^0 l^{\pm}, \\
\widetilde{\chi}_1^{\pm} &\rightarrow \tilde{\nu}_L l^{\pm} \rightarrow \widetilde{\chi}_1^0 \nu, \\
\widetilde{\chi}_1^{\pm} &\rightarrow \widetilde{\chi}_1^0 W^{\pm} \rightarrow l^{\pm}\nu.
\end{aligned}$$

The branching ratios for these decay channels have been presented in Ref.[5], as a function of the SUSY breaking terms (m_0, $m_{1/2}$). It was shown that in different regions these different leptonic decay modes are complementary. Three body decays are open in the region $m_{1/2} \lesssim 200$ GeV and $m_{1/2} \gtrsim 0.5m_0$, whereas in the rest of the parameter space the two body decays are dominant. Leptonic dacays of $\widetilde{\chi}_2^0$ which give two isolated leptons and missing energy ($\widetilde{\chi}_1^0$) have been also studied. It was emphasized that, like the $\widetilde{\chi}_1^{\pm}$ case, the three body decay branching ratios are sizable at $m_{1/2} \lesssim 200$ GeV and $m_{1/2} \gtrsim 0.5m_0$. In hadronic collisions, the charginos and neutralinos can be produced directly via 21 different reactions or through the cascade decays of strongly interacting sparticles.

Gluinos and squarks

In mSUGRA the gluino mass is determined by $m_{1/2}$: $M_{\widetilde{g}} \approx 2.5m_{1/2}$. The five squark flavours $(\widetilde{d}, \widetilde{u}, \widetilde{s}, \widetilde{c}, \widetilde{b})$ with their left and right chiral states are assumed to be mass degenerate, giving ten degenerate squark flavours with: $M_{\widetilde{q}} \approx \sqrt{m_0^2 + 6m_{1/2}^2}$. The $\widetilde{t}_L$ and $\widetilde{t}_R$ are treated differently since they have a large mixing which leads to two mass eigenstates $\widetilde{t}_1$ and $\widetilde{t}_2$, with $\widetilde{t}_1$ significantly lighter than all other $\widetilde{q}$ states, which requires a separate discussion in both cases of production and decay.

As we said, at LHC the total SUSY particle production cross section is dominated by strongly interacting gluinos and squarks, which through their cascade decays can produce many jets and leptons with missing energy. For instance squark and gluinos can have strong decays [5],

$$\widetilde{q}_{L,R} \quad \rightarrow \quad q\widetilde{g},\ q\bar{\widetilde{q}}_{L,R},\ q\widetilde{q}_{L,R},$$

or they can have weak decays

$$\widetilde{q}_{L,R} \rightarrow q\widetilde{\chi}_i^0, \quad \widetilde{q}_L \rightarrow q'\widetilde{\chi}_i^{\pm}, \quad \widetilde{g} \rightarrow q\bar{q}\widetilde{\chi}_i^0,\ q\bar{q}'\widetilde{\chi}_i^{\pm},\ g\widetilde{\chi}_i^0,$$

where the strong deacays are dominant if they are kinematically allowed. We present in Table 5 the total cross sections for pair productions of SUSY particles at the five LHC points. If the lighter stop eigenstate $\widetilde{t}_1$ is the lightest visible SUSY particle, it decays into $c\widetilde{\chi}_1^0$ or $bW^{\pm}\widetilde{\chi}_1^0$.

As can be seen from Table 4, the gluinos and in particular squarks can have large branching ratios for decays into charginos and higher neutralinos, which may then decay

TABLE 4. Cross sections for production of supersymmetric particles, at the five LHC SUGRA points [12].

Point	1	2	3	4	5
$\tilde{g}\tilde{g}$	258	259	437189	10877	1751
$\tilde{q}\tilde{q}$	363	337	103059	455	2379
$\tilde{q}\bar{\tilde{q}}$	686	672	73769	909	2820
$\tilde{b}_i\bar{\tilde{b}}_i$	54	34	18442	57	297
$\tilde{t}_i\bar{\tilde{t}}_i$	150	95	18985	293	701
$\tilde{g}\tilde{q}$	1486	1444	642765	8259	8306
$\tilde{\chi}_i^{\pm}\tilde{\chi}_j^{\mp}$	66	75	7865	1108	242
$\tilde{\chi}_i^0\tilde{\chi}_j^0$	6	16	814	110	18
$\tilde{\chi}_i^{\pm}\tilde{\chi}_j^0$	138	146	13832	3532	521
$\tilde{l}_i\bar{\tilde{l}}_i, \tilde{\nu}_i\bar{\tilde{\nu}}_i$	9	13	542	-	253

in several steps. Both the ATLAS and the CMS detectors at LHC are designed to clearly identify and analyse events which exhibit one or more of these characteristic features.

CONCLUSIONS

We have analyzed the discovery potential for the supersymmetric particles (squarks, gluinos, sleptons, charginos, neutralinos, and lightest Higgs) at the LHC, which will start operating in 2006, producing energies about 1000 GeV. We have shown that if supersymmetry exists at the electroweak scale, then these particles could be easily observed at the LHC. As a particular example, we have assumed the mSUGRA scenario where the LSP is stable and which is the most interesting candidate for dark matter. In this case the discovery of supersymmetry should be relatively straightforward.

REFERENCES

1. A review of the Standard Model can be found in P. Kalmus, Contemp. Phys. **41**, 129 (2000).
2. For introduction to superesymmetry, see G. Kane, Contemp. Phys. **41**, 359 (2000).
3. J. R. Ellis, S. Kelley and D. V. Nanopoulos, Phys. Lett. B **260**, 131 (1991); U. Amaldi, W. de Boer and H. Furstenau, Phys. Lett. B **260**, 447 (1991).
4. H. E. Haber and G. L. Kane, Phys. Rept. **117**, 75 (1985); J. R. Ellis, arXiv:hep-ph/0203114; S. P. Martin, arXiv:hep-ph/9709356; J. R. Ellis, arXiv:hep-ph/9812235; H. Murayama, arXiv:hep-ph/0002232;
5. S. Abdullin *et al.* [CMS Collaboration], J. Phys. G **28**, 469 (2002) [arXiv:hep-ph/9806366].
6. L. Girardello and M. T. Grisaru, Nucl. Phys. B **194**, 65 (1982).
7. F. E. Paige, arXiv:hep-ph/9801254.
8. K. Hagiwara et al., Phys. Rev. D66, 010001 (2002).
9. S. Chen *et al.* [CLEO Collaboration], Phys. Rev. Lett. **87**, 251807 (2001)
10. S. Khalil and C. Munoz, Contemp. Phys. **43**, 51 (2002).
11. A. Masiero and O. Vives, arXiv:hep-ph/0003133.
12. A. Bartl *et al.*, LBL-39413 *Presented at 1996 DPF / DPB Summer Study on New Directions for High-Energy Physics (Snowmass 96), Snowmass, CO, 25 Jun - 12 Jul 1996.*

Precision Tests of Electroweak Interactions

Arif Akhundov

Institute of Physics, Azerbaijan Academy of Sciences, 370143 Baku, Azerbaijan
and
Departamento de Física Teórica and IFIC, Universidad de Valencia-CSIC, 46100 Valencia, Spain

Abstract. The status of the precision tests of the electroweak interactions is reviewed in this paper. An emphasis is put on the Standard Model analysis based on measurements at LEP/SLC and the Tevatron. The results of the measurements of the electroweak mixing angle in the NuTeV experiment and the future prospects are discussed.

Keywords: Electroweak interactions, Precision tests, Electroweak radiative corrections
PACS: 12.15.Lk, 13.66.Jn, 14.40.Ks

1. INTRODUCTION

The unification of the electromagnetic and weak interactions in 1968 [1], the discoveries of the neutral currents in 1973 [2], of the charm quark in 1974 [3], of the W and Z bosons in 1983 [4] were very successful steps for the theory of the ElectroWeak (EW) interactions, the Standard Model (SM) [1, 5]. After the discoveries of the top quark in 1995 [6] and the tau neutrino in 2000 [7] the electroweak SM became the commonly accepted theory of the fundamental electroweak interactions. It is a gauge invariant quantum field theory based on the symmetry group $SU(2) \times U(1)$, which is spontaneously broken by the Higgs mechanism. The renormalizability of the SM [8] allows us to make precise predictions for measurable quantities at higher orders of the perturbative expansion, in terms of a few input parameters. The higher-order terms, Radiative Corrections (RC) or quantum corrections, contain the self-coupling of the vector bosons as well their interactions with the Higgs field and the top quark. Their calculation provides the theoretical basis for the EW precision tests.

In the last thirty five years in High Energy Physics two distinct and complementary strategies have been used for gaining new understanding of the Nature:

- The Direct Discovery of the New Phenomena at High Energy accelerators
- The Precision Measurements of the Known Phenomena at existing accelerators with high Luminosity

The excellent example is the ratio (Fig. 1)

$$\mathrm{R}_{e^+e^-} \equiv \frac{\sigma(e^+e^- \to \text{hadrons})}{\sigma(e^+e^- \to \mu^+\mu^-)}. \tag{1}$$

Both strategies have worked very well for the studies of the EW interactions first at SPS and LEP1/SLC, and later at LEP2/SLD and the Tevatron, and in the future such interplay of the proton and electron colliders will be applied for the LHC and ILC/CLIC.

CP1006, *High Energy Physics and Applications*, edited by J. Ellis, S. Nasri, and E. Malkawi

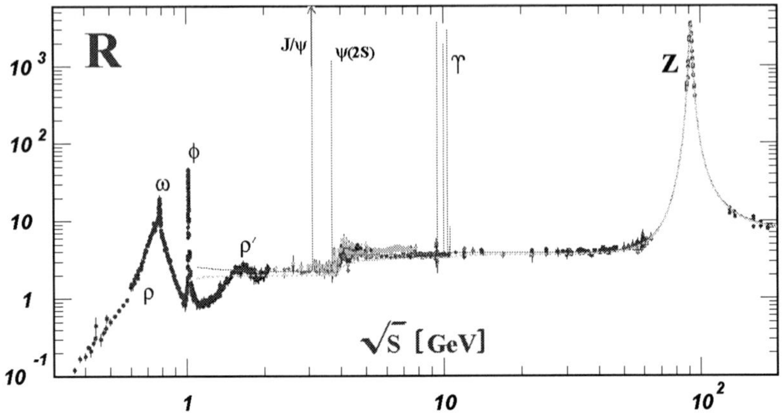

FIGURE 1. World data on the ratio $R_{e^+e^-}$ [9].

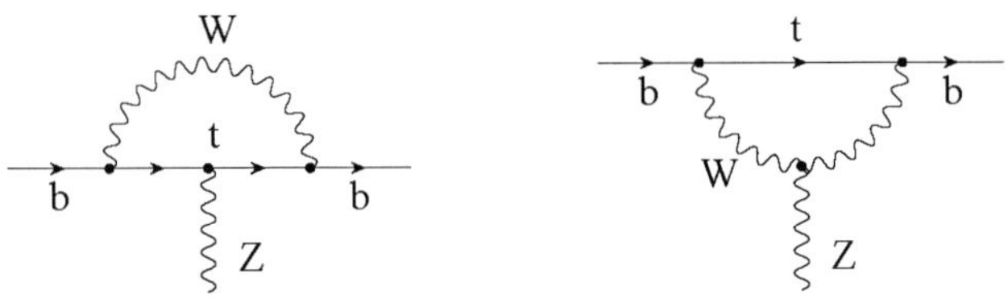

FIGURE 2. One-loop corrections to the $Zb\bar{b}$ vertex, involving a virtual top quark [12].

The second strategy has always demanded from the theory the prediction of physical quantities with high precision, *i.e.* at the level of quantum corrections. For the interpretation of the precision experiments the RC play the crucial role.

2. ELECTROWEAK RADIATIVE CORRECTIONS

In the context of the SM any electroweak process can be computed at tree level from α (the fine structure constant measured at values of Q^2 close to zero), M_W (the W-boson mass), M_Z (the Z-boson mass), and V_{jk} (the Cabbibo-Kobayashi-Maskawa flavor mixing matrix elements) [10].

When higher order corrections are included, any observable can be predicted using the on-shell renormalization scheme [11] as a function of:

$$O_i = f_i(\alpha, \alpha_s, M_W, M_Z, M_H, m_f, V_{jk}), \tag{2}$$

where the effects of heavy particles do not decouple, and there is the sensitivity to the top mass m_t [12] and to less extend to the Higgs mass M_H [13]. Since the discovery of the presence of hard m_t^2 corrections to the Zbb vertex (see Fig. 2) [12, 14, 15, 16] the calculation of the EW RC has been theoretically well established and many higher-order

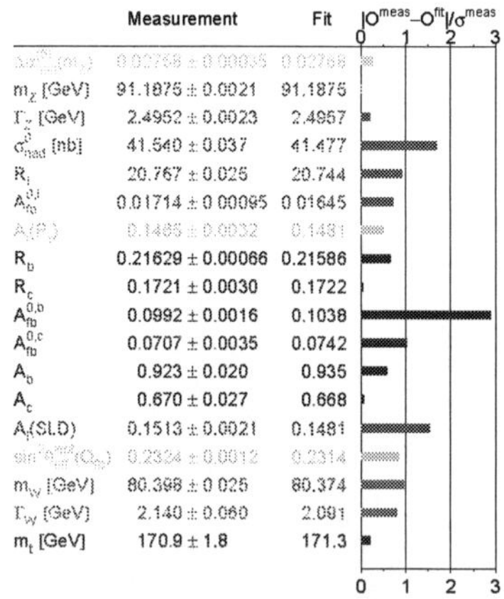

FIGURE 3. Comparison between the measurements included in the combined analysis of the SM and the results from the global EW fit [17, 18].

contributions of the radiative corrections have become available over past decades to improve and stabilize the SM predictions.

3. TESTS OF EW INTERACTIONS AT LEP/SLC AND TEVATRON

The experimental data for testing of the EW theory have achieved an impressive accuracy. After taking the measured Z mass, besides α and G_μ (the Fermi constant measured in the muon decay), for completion of the input,

$$M_Z = (91.1875 \pm 0.0021)\,\mathrm{GeV} \quad [17, 18], \tag{3}$$

$$G_\mu = (1.166371 \pm 0.000006) \cdot 10^{-5}\,\mathrm{GeV}^{-2} \quad [19], \tag{4}$$

$$\alpha^{-1} = 137.035999710 \pm 0.000000096 \quad [20] \tag{5}$$

each other precision observable provides a test of the electroweak theory (Fig. 3). The predictions are calculated with computer programs ZFITTER [21] and TOPAZ0 [22], which incorporate state-of-the-art calculations of the EW, QED and QCD radiative corrections.

Theoretical predictions of the SM depend on the mass of the top quark and of the as yet experimentally unknown Higgs boson through the virtual presence of these particles in the loops. As a consequence, precision data can be used to pin down the allowed range of the mass parameters. This is shown in Fig. 4, which compares the information on M_W and m_t obtained at LEP1 and SLD, with the direct measurements performed at LEP2 and the Tevatron.

The measured at Tevatron mass $m_t = 170.9 \pm 1.9$ [23] agrees better than 10 % with the value predicted within the SM on the basis of the precision EW measurements.

Taking all direct and indirect data into account, one obtains the pillar of the precision electroweak physics [24]: the best constraints on the possible mass M_H of unseen Higgs.

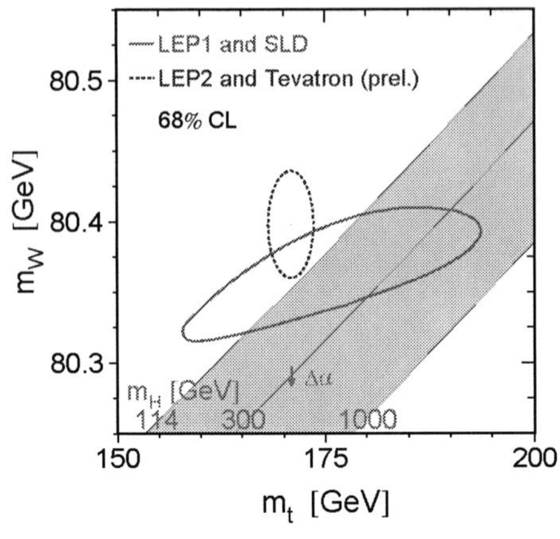

FIGURE 4. Comparison of the direct measurements of M_W and m_t at LEP2/Tevatron with the indirect determination through electroweak radiative corrections at LEP1/SLD. Also shown in the SM relationship for the masses as function of M_H [17, 18].

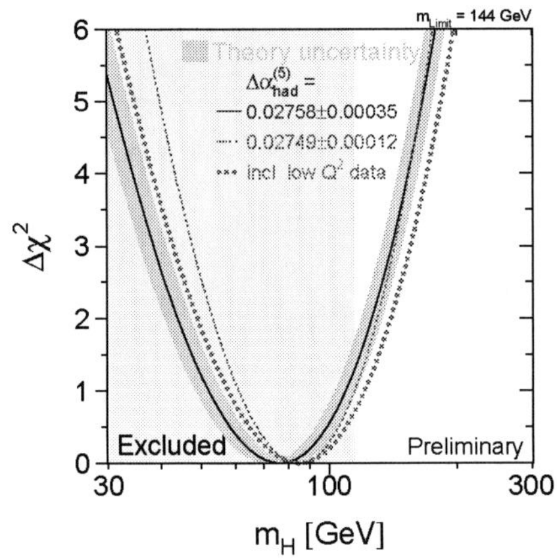

FIGURE 5. $\Delta\chi^2 = \chi^2 - \chi^2_{min}$ versus M_H, from the global fit to the EW data. The vertical band indicates the 95% exclusion limit from direct searches [17, 18].

The global electroweak fit results in the $\Delta\chi^2 = \chi^2 - \chi^2_{min}$ curve shown in Fig. 5. The lower limit on M_H obtained from direct searches is close to the point of minimum χ^2. At 95% C.L., one gets [17, 18]

$$114.4\,\text{GeV} < M_H < 144\,\text{GeV}.$$

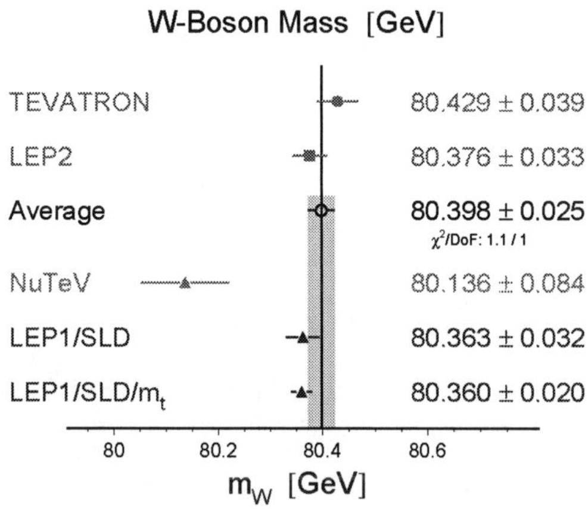

FIGURE 6. The results of the direct measurements of M_W at LEP2/Tevatron are compared with the indirect determinations at LEP1/SLD and in the NuTeV experiment [17, 18, 25].

4. SM ANALYSIS AND NUTEV EXPERIMENT

In the on-shell scheme [11] the three-level formula $\sin^2\theta_W = 1 - \frac{M_W^2}{M_Z^2}$ is a definition of the renormalized $\sin^2\theta_W$ to all orders in perturbation theory, *i.e.*,

$$\sin^2\theta_W{}^{\text{on-shell}} \equiv s_W{}^2 = 1 - \frac{M_W^2}{M_Z^2}. \tag{6}$$

A precise determination of the on-shell EW mixing angle has been performed by the NuTeV collaboration [25] for the first time through the measurements of the Pashos-Wolfenstein ratio:

$$R^- \equiv \frac{\sigma(\nu_\mu N \to \nu_\mu X) - \sigma(\bar{\nu}_\mu N \to \bar{\nu}_\mu X)}{\sigma(\nu_\mu N \to \mu^- X) - \sigma(\bar{\nu}_\mu N \to \mu^+ X)} \tag{7}$$

from deep inelastic neutrino scattering on isoscalar targets. The NuTeV collaboration finds $s_W{}^2 = 0.2277 \pm 0.0016$ which is 3.0 σ higher than the SM predictions.

From this experimental value one obtains the mass of M_W boson [25]

$$M_W = 80.14 \pm 0.08\ \text{GeV} \tag{8}$$

which is smaller than other measurements of M_W at LEP/SLD and the Tevatron (see Fig. 6). The NuTeV result should be considered as preliminary until a reanalysis of data will be completed including all experimental and theoretical information.

5. CONCLUSIONS AND FUTURE PROSPECTS

Apart from the still missing Higgs boson, the SM provides an elegant theoretical framework for the description of the known experimental facts in Particle Physics. The SM has been impressively confirmed by successful collider experiments at the particle accelerators LEP, SLC and Tevatron during the last fifteen years.

Future colliders like the upcoming LHC or an ILC/CLIC offer great prospects, and in turn represent a great challenge for theory to provide even more precise calculations. Accurate predictions are necessary not only to increase the level of precision of SM tests, but also to study the indirect effects of possible new particles.

ACKNOWLEDGMENTS

I would like to thank the organizers of the UAE-CERN Workshop for kind invitation to give a plenary talk and for the hospitality at Al-Ain. I am grateful to J. Vidal for support and to A. Shiekh for remarks.

REFERENCES

1. S. Weinberg, *Phys. Rev. Lett.* **19**, 1264 (1967). A. Salam, in *Elementary Particle Theory*, ed. N. Svartholm (Almquist and Wiksells, Stockholm, 1969), p. 367.
2. F.J. Hasert *et al.*, *Phys. Lett.* **B46**, 121 (1973); *Phys. Lett.* **B46**, 138 (1973).
3. J.J. Aubert *et al.*, *Phys. Rev. Lett.* **33**, 1404 (1974). J.E. Augustin *et al.*, *ibid.* **33**, 1406 (1974).
4. UA1 Collaboration, G. Arnison *et al.*, *Phys. Lett.* **B122**, 103 (1983); *Phys. Lett.* **B126**, 398 (1983). UA2 Collaboration, B. Banner *et al.*, *Phys. Lett.* **B122**, 476 (1983); *Phys. Lett.* **B129**, 130 (1983).
5. S.L. Glashow, *Nucl. Phys.* **22**, 579 (1961). S.L. Glashow, J. Iliopoulos and L. Maiani, *Phys. Rev.* **D2**, 1285 (1970).
6. CDF Collaboration, F. Abe *et al.*, *Phys. Rev. Lett.* **74**, 2626 (1995). D0 Collaboration, S. Abachi *et al.*, *Phys. Rev. Lett.* **74**, 2632 (1995).
7. DONUT Collaboration, K. Kodama *et al.*, *Phys. Lett.* **B504**, 218 (2001).
8. G. 't Hooft, *Nucl. Phys.* **B33**, 173 (1971).
9. W.-M. Yao *et al.*, *Review of Particle Physics, J. Phys.* **G33**, 1 (2006).
10. N. Cabibbo, *Phys. Rev. Lett.* **10**, 531 (1963). M. Kobayashi, K. Maskawa, *Progr. Theor. Phys.* **42**, 652 (1973).
11. A. Sirlin, *Phys. Rev.* **D22**, 971 (1980); *Phys. Rev.* **D29**, 89 (1984).
12. A. A. Akhundov, D. Yu. Bardin and T. Riemann, *Nucl. Phys.* **B276**, 1 (1986).
13. A. A. Akhundov, D. Yu. Bardin and T. Riemann, *Phys. Lett.* **B166**, 111 (1986).
14. W. Beenakker and W. Hollik, *Z. Phys.* **C40**, 141 (1988).
15. B.W. Lynn and R.G. Stuart, *Phys. Lett.* **B252**, 676 (1990).
16. J. Bernabeu, A. Pich and A. Santamaria, *Nucl. Phys.* **B363**, 326 (1991).
17. The LEP Collaborations ALEPH, DELPHI, L3 and OPAL and the LEP Electroweak Working Group, arXiv:hep-ex/0612034; http://www.cern.ch/LEPEWWG/.
18. The ALEPH, DELPHI, L3, OPAL and SLD Collaborations, the LEP Electroweak Working Group and the SLD Electroweak and Heavy Flavour Groups, *Phys. Rept.* **427**, 257 (2006).
19. MuLan Collaboration, arXiv:hep-ex/0704.1981.
20. G. Gabrielse *et al.*, *Phys. Rev. Lett.* **97**, 030802 (2006).
21. D.Yu. Bardin *et al.*, *Z. Phys.* **C44**, 493 (1989); *Comp. Phys. Comm.* **59**, 303 (1990); *Nucl. Phys.* **B351**, 1 (1991); *Comp. Phys. Comm.* **133**, 229 (2001).
22. G. Montagna *et al.*, *Nucl. Phys.* **B401**, 3 (1993); *Comp. Phys. Comm.* **76**, 328 (1993); *Comp. Phys. Comm.* **93**, 120 (1996); *Comp. Phys. Comm.* **117**, 278 (1999).
23. CDF and D0 Collaborations, arXiv:hep-ex/0703034.
24. John Ellis, *Plenary talk at International Conference on High Energy Physics (EPS-HEP2007)*, Manchester, England, 19-25 July 2007; arXiv:hep-ph/0710.5590.
25. NuTeV Collaboration, G.P. Zeller *et al.*, *Phys. Rev. Lett.* **88**, 091802 (2002) [Erratum-ibid. **90**, 239902 (2003)]

QCD Jets from I.S.R. to L.H.C. Energies

Hariri Saadeddine

Lebanese University, Faculty of Sciences I, Lebanon

Abstract. In this article, a Lorentz invariant technique of reconstructing jets of hadrons produced in proton-proton deep inelastic scattering at LHC energies is described. The analysis is able to handle events with two, three and four jets with equal efficiency and precision. The resolution on the jet axis and energy is relatively high. The analysis allows a good measurement of the jet cross section, jet fragmentation function and of the proton structure function.

Keywords: Transverse jets, LHC, energy flow, proton-proton.

INTRODUCTION

Proton-proton deep inelastic scattering occurs when two partons, each originating from one of the protons and carrying the fractions x_1, x_2 of the beam energy, undergo hard scattering with a large transfer of four momentum. The outgoing partons are emitted at wide angles with respect to the beam axis. After hadronization, these partons generate two jets with high transverse momemntum K_t that will be called abusively transverse jets. The physics of these transverse jets is the main object of this article.

Initialization

The fragments of the scattered protons decay into two energetic jets emitted along the beam axis. These longitudinal jets produce two peaks, in the rapidity distribution dE/dy of the hadronic energy, located near the edges of the rapidity interval used by the final state hadrons. From the experimental distribution dE/dy, the widest rapidity interval $[-y_0, y_0]$ that excludes these two peaks of energy flow can be determined. This defines the rapidity cut $|y|\leq y_0$ that we introduce to remove most of the energy generated by the longitudinal jets. Moreover, the particles of these jets are produced with a small transverse momentum q_T relative to the beam axis. By cutting out particles with low transverse momentum, $q_T\leq$ 0.5 Gev or so, the contribution of the longitudinal jets to the total observed energy E_{vis} of an event is minimized. The q_T cut can be optimized using the measured spectrum $d\sigma/dq_T$ of the hadronic transverse momentum. From now on, the energy E_{vis} will represent the sum of the energies of all particles surviving the q_T cut and the y_0 cut.

The position of a vector $\mathbf{q_T}$ in the transverse plane is referenced by the angular coordinate Φ. Particles of the longitudinal jets passing the (y_0, q_T)

CP1006, *High Energy Physics and Applications*, edited by J. Ellis, S. Nasri, and E. Malkawi

cuts generate a hadronic background energy. The distribution of this energy in the (y, Φ) plane is flat and is characterized by its low height. A high energy transverse jet is seen as a large amount of hadronic energy concentrated within a small rectangle in the (y,Φ) plane. This is exactly the genius achievement of the (UA1, UA2) collaborations who published the angular distribution in the (Φ,θ) plane of the transverse hadronic energy of events with two and three transverse jets produced in proton-antiproton hard scattering at the S.P.S. energies.

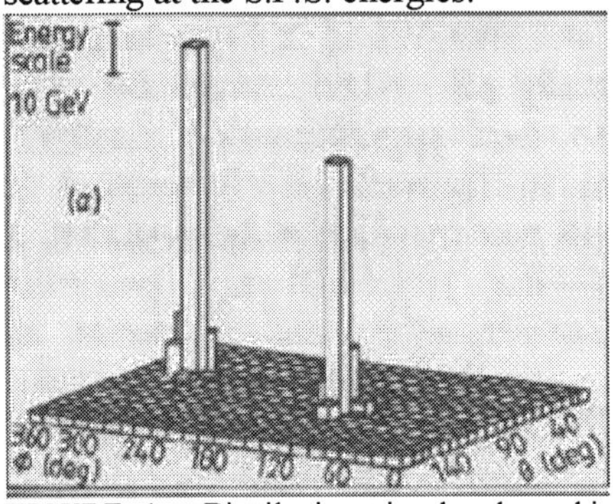

FIGURE 1. Distributions in the theta-phi plane of the transverse energy of two jet events from the UA2 collaboration (Di Lella 1985). Each bin represents a cell of the UA2 calorimeter.

In our approach, the energy E_{vis} is spread over a rectangle of dimensions $2y_0 \times 360°$ and the resulting average density of energy per degree and per unit of rapidity is order 1 Gev at L.H.C. energies outside a transverse jet and it is of the order of 70 within a transverse jet of 70 Gev. We do characterize the transverse jets by the high density of energy $dE_{vis}/d\Phi dy$ within the jet. The results of the (UA1,UA2) experiments represent a strong support to this picture of a jet as a high density flux of energy of limited angular aperture.

To initialize the analysis, we choose to start with a clean sample of high energy transverse jets emitted at wide angles with respect to the beam axis. This can be easily achieved by requiring that the fraction of E_{vis} generated by particles with $|y| \leq (y_0/2)$ to be of the order of 0.6 to 0.7 of the beam energy. We take the y-axis along the beam axis and the z-axis to be the axis of the transverse thrust calculated using the q_T vectors of the particles. The unit vector $\mathbf{\hat{u}_z}$ of the thrust satisfies the condition: $T_t = \sum|\mathbf{q}_T.\mathbf{\hat{u}}_z|/\sum|\mathbf{q}_T|$= maximum. The axes of the transverse jets are in the yz -plane. The xy -plane of our Cartesian frame divides the event into two hemispheres and two equivalent sub-events with one transverse jet per hemisphere. In each of the hemispheres, the angular coordinate Φ is zero on the z-axis and it runs from -90° to +90°. We take E_h to be the hadronic energy of a hemisphere and we calculate the scaled energy flow$(100/ E_h)dE_h/d|\Phi|$ for all sub-events. This distribution calculates in percent the density of energy per unit of angle Φ inside and outside the jet. The angular size Φ_0 of the jet is the full width of the peak of energy flow around the point Φ=0. All the particles of the jets do satisfy the requirements $|\Phi| \leq \Phi_0$. The angle Φ_0 also determines the rapidity width of a jet as $\Delta y= -Ln(\tan(\Phi_0/2))$. Just by precaution, we take $Y_0= 2\ \Delta y$ and we set E'_h to be the energy of all particles with $|\ \Phi|\leq \Phi_0$ and we calculate the energy $E(y_j)$ of all particles with $|\Phi|\leq \Phi_0$ and $|y-y_j|\leq Y_0$. The rapidity position of the jet is such that $E(y_j)$ is maximum. The pseudo rapidity distribution $(1/\ E'_h)d\ E'_h/dy'$ with $y'=|y-y_j|$ is strongly peaked at $y'=0$ and it keeps a sizeable height within the segment $y'\leq Y_0$. To obtain a statistically significant measure of Y_0, we plot this pseudo rapidity distribution of E'_h for all selected sub-events and we redefine Y_0 as the full width of the obtained spectrum. With this experimental determination of the parameters (Φ_0, Y_0), we terminate the initialization of the analysis.

Application to L.H.C. Jets

The full geometrical acceptance for jet production is defined in the (Φ,y)-plane by the rectangle $\{0<\Phi<360°;-y_0<y<y_0\}$. The average density of energy per degree and per unit of rapidity $f_0=<dE_{vis}/d\Phi dy>= E_{vis}/(2y_0\times 360°)$ is order 1 Gev at L.H.C. energies. In contrast to that, the average density of energy $E_j/(2\Phi_0\ \times 2Y_0)$ within a jet, of energy E_j = 70 Gev or above, is order E_j. Consequently , the height of the peak of energy associated with a jet is much higher than the density of energy outside the jets (See Fig-1). The function $f(\Phi,y)= E(\Phi,y)/\ (2\Phi_0\times 2Y_0)$ associated with the point(Φ,y) expresses the ratio of the energy of all particles within the rectangle $\{\ (\Phi-\ \Phi_0,\ \Phi+\Phi_0);\ (y-\ Y_0,\ y_+\ Y_0)\}$divided by the rectangle area. A maximum $f(\Phi_j,y_j)$ corresponds either to a local fluctuation of energy with a value comparable to f_0 or to the position(Φ_j,y_j) of a jet axis if its value is two or three orders of magnitude larger than f_0. This is exactly what is visualized in FIG-1. Therefore, the sharp peaks of the function $f(\Phi,y)$ corresponds to the location of energetic jets of hadrons. In this way, events with two, three and four jets of hadrons can be easily reconstructed with a very high resolution on the position of the jet axis.

Conclusions

The concept of a jet as a high intensity flux of energy of limited angular aperture is deeply enrooted in the history of physics. A statistically significant signal for three jet production in electron-positron annihilations was published by the MARKJ –Collaboration in the form of a density of energy per unit angle calculated in the event plane with the zero angle position taken along the axis of the most energetic jet, [2]. This is one of the most important achievements of the PETRA accelerator.

In a QCD based Monte-Carlo program, the jet production proceeds through the hadronization of parton showers emitted by a highly energetic final state parton. The momentum structure and angular width of a jet are fixed by these parton showers of the final state, [3]. In this analysis, a deep test of the jet production mechanism is achieved through the precise measurement of the distribution $(1/E_j)dE_j/dyd\Phi$ of the jet energy E_j around the jet axis. Moreover, the jet energy and jet axis are reconstructed with a high precision. This allows a good measurement of the initial momemnta x_1, x_2 of the colliding partons and of the Q^2 of the scattering process. This leads us to a measurement of the proton structure function at high Q^2 and especially to a determination of the gluon contents of a high energy proton.

ACKNOWLEDGMENTS

I would like to thank the Dean and the Professors of the Faculty of Sciences at the UAEU for the excellent of the UAE-CERN workshop and their generous hospitality. At the same time, I would like to thank Professors J. Ellis, E. Tsesmelis, C. Fabjan and B. Peyaud for the illuminating contributions and fruitful discussions.

REFERENCES

1. L. Di Lella, , *Ann. Rev. Nucl. Sci.* ***35***; 1985, p. 107.
2. D. P. Barber et al., Discovery of Three Jet Events and a Test of Quantum Chromodynamics at PETRA, *Phys. Review Letters* **83**, No. 12, p.832 (17 Sep. 1979).
3. S. Hariri, A High Resolution Jet Analysis for LEP, *LAPP-EXP-92.03* (November 1992).

Strangeness in Nucleon

H.B. Benaoum [1]

Prince Mohammad University, Al-Khobar 31952, Saudi Arabia
Syracuse University, Syracuse, NY 13244, USA
Thomas Jefferson National Accelerator Laboratory, Newport News, VA 23606, USA

Abstract. Results of the parity violating asymmetry A_{PV} for longitudinally polarized 3 GeV electrons from both hydrogen and helium cryogenic targets, at small scatteting angle $\theta_{lab} \sim 6°$ are presented. The asymmetry for hydrogen is a function of a linear combination of G_E^s and G_M^s, the strange quark contributions to the electric and magnetic form factors of the nucleon respectively, and that for ^{4}He is a function solely of G_E^s. The combination of the two results therefore allows G_E^s and G_M^s to be separately determined.

Keywords: Strangeness, nucleon, form factors, electroweak interations, parity violation
PACS: 12.10.Dm, 12.15.-y, 14.20.Dh, 21.00.00, 24.80.+y

INTRODUCTION

Our universe is made up of dark energy, dark matter and ordinary baryonic matter. The nature of the dark energy and matter is the biggest mystery in cosmology today. However the ordinary baryonic matter is made up largely of baryons. To understand the latter we need to learn more about the internal structure of the nucleon.

There is currently enormous interest in the determination of the strangeness content of the nucleon which is crucial in the understanding of quantum chromodynamics (QCD) in the non-perturbative regime. In this regime, the QCD coupling constant is large and in addition to the valence quarks (up and down quarks) there is a sea of gluons and $q\bar{q}$ pairs that plays an important role at these distance scales.

Because the strange quark mass is comparable to the QCD mass scale Λ_{QCD}, it is reasonable to assume that it might have non-negligible contributions to the nucleons properties. Possibilities of the strange quark contributions to nucleon as $s\bar{s}$ sea were suggested by Kaplan and Manohar [1]. They showed how to probe neutral current with the help of scattering experiments to extract information on the strange matrix elements for nucleon.

The parity-violating (PV) electroweak interaction in the electron-nucleon elastic scattering offers a new opportunity to study the contribution of the strange quarks to the electroweak structure of the nucleon [2, 3, 4, 5, 6].

---

[1] email : hbenaoum@pmu.edu.sa and hbenaoum@jlab.org

CP1006, *High Energy Physics and Applications*, edited by J. Ellis, S. Nasri, and E. Malkawi

PARITY-VIOLATING ELECTRON SCATTERING

In parity-violating electron scattering, the asymmetry due to the interference of the electromagnetic and weak currents, is given by :

$$A_{PV} = \frac{\sigma_R - \sigma_L}{\sigma_R + \sigma_L} \tag{1}$$

where $\sigma_{R(L)}$ is the differential cross section for elastic scattering of right and left handed longitudinally polarized electrons from unpolarized target.

This asymmetry can be written in terms of electromagnetic nucleon N (proton or neutron) form factors $G_E^{(\gamma,N)}$ and $G_M^{(\gamma,N)}$ and the electric $G_E^{(Z,N)}$, magnetic $G_M^{(Z,N)}$ and axial G_A^{eN} weak form factors.

These form factors can be re-expressed in terms of its quark contributions as :

$$G_{E,M}^{(\gamma,N)} = \sum_{q=u,d,s} Q_q G_{E,M}^{(q,N)}, \quad G_{E,M}^{(Z,N)} = \sum_{q=u,d,s} C_V^q G_{E,M}^{(q,N)}, \quad G_A^{eN} = \sum_{q=u,d,s} C_A^q G_A^{(q,N)} \tag{2}$$

where C_V^q and C_A^q are the vector and axial coupling to quarks.

Charge symmetry approximation between proton and neutron implies that :

$$G_{E,M}^{(u,p)} = G_{E,M}^{(d,n)}, \quad G_{E,M}^{(s,p)} = G_{E,M}^{(s,n)} \tag{3}$$

It is possible to express the electric and magnetic quark form factors as a function of the electromagnetic and weak form factors of the proton and the neutron. In particular, the extraction of the strange form factor requires the measurement of the electromagnetic and weak form factors through the relation :

$$G_{E,M}^s = (1 - 4\sin^2\theta_W)\, G_{E,M}^{(\gamma,n)} - G_{E,M}^{(\gamma,n)} - G_{E,M}^{(Z,p)} \tag{4}$$

Parity violation in elastic scattering from the ^4He nucleus is directly sensitive to G_E^s,

$$A_{PV}^{He} = \frac{G_E Q^2}{4\pi\alpha\sqrt{2}} \left(4\sin^2\theta_W + \frac{2G_E^s}{G_E^{(\gamma,p)} + G_E^{(\gamma,n)}} \right) \tag{5}$$

whereas parity violation in elastic scattering from the proton is sensitive to the linear combination of G_E^s and G_M^s,

$$\begin{aligned} A_{PV}^H = & -\frac{G_F Q^2}{4\pi\alpha\sqrt{2}} \left(1 - 4\sin^2\theta_W - \frac{\varepsilon G_E^{(\gamma,p)}(G_E^{(\gamma,n)} + G_E^s) + \tau G_M^{(\gamma,p)}(G_M^{(\gamma,n)} + G_M^s)}{\varepsilon G_E^{(\gamma,p)\,2} + \tau G_M^{(\gamma,p)\,2}} \right. \\ & \left. - \frac{(1 - 4\sin^2\theta_W)\varepsilon' G_M^{(\gamma,p)} G_A^{(Z,p)}}{\varepsilon G_E^{(\gamma,p)\,2} + \tau G_M^{(\gamma,p)\,2}} \right) \end{aligned} \tag{6}$$

Thus by combining measurement from the proton and from the helium, we can separate G_E^s and G_M^s.

HAPPEX II MEASUREMENT

The Happex measurement was carried out at Thomas Jefferson National Accelerator Facility (JLab) [9, 10, 11, 12]. A 35 to 55 μA longitudinally polarized electrons beam with an energy of 3 GeV was incident on 20 cm long cryogenic targets. Elastically scattered electrons with $\theta_{lab} \sim 6^\circ$ was focused by two identical 3.7 mr spectrometer systems consisting of the Hall A superconducting septum magnets plus high resolution spectromters (HRS).

The scattered electrons were intercepted by two identical detectors, located in the focal plane of each spectrometer. These detectors were made of sandwiched layers of absorber (brass plate) and optical material (fuzed quartz). The Čerenkov light generated by electromagnetic shower was transported by the quartz to one end of the detector and collected by photomultiplier tubes (PMTs).

The polarized electron beam originates from a GaAs photocathode excited by circularly polarized laser light. High polarization of the extracted electron beam is achieved using superlattice of doped GaAs semiconductor layers grown by molecular beam epitaxy (MBE). The beam helicity is first determined by an electronics module located near the polarized source and set every 33 ms. The helicity sequence consists of pair of consecutive windows with opposite helicity.

The response of beam monitors and electron detectors PMTs is integrated over each window and then digitized by analog to digital converters (ADCs).

The HRS pair has sufficient momentum resolution to separate cleanly the elastic electrons from the inelastic one and allows us to spatially position the integrating detectors in order to maximize signal from elastic peak.

Auxilliary measurements to determine the amount of background and Q^2 were performed using counting mode method at low rate.

Continuous monitoring of the beam polarization was done by the Compton polarimeter and invasive measurements of the beam polarization were performed at low current by the Moller and Mott polarimeters. Good agreement in measuring the beam polarization by the three polarimeters was noticed.

The raw asymmetry A_{raw} of the integrated detector flux for each pair was calculated by taking left-right integrated signal and normalizing it to the beam intensity. The raw asymmetry was then corrected for the helicity-correlated beam asymmetries $A_{beam} = \Sigma_{i=1}^{5} \beta_i \Delta x_i$, according to :

$$A_{corr} = A_{raw} - \Delta A \tag{7}$$

where β_i is the normalized detector sensitivity to the beam position at the ith BPM. The final corrected asymmetries are :

$$\begin{aligned} A_{PV}^{He} &= +5.25 \pm 0.19_{stat} \pm 0.05_{syst}\ ppm \qquad \text{at } Q^2 = 0.077\ GeV^2 \\ A_{PV}^{H} &= -1.42 \pm 0.11_{stat} \pm 0.02_{syst}\ ppm \qquad \text{at } Q^2 = 0.109\ Gev^2 \end{aligned} \tag{8}$$

The physical asymmetry A_{phys} is formed by correcting it from beam polarization P_b, background fractions f_i with asymmetries A_i and finite kinematic acceptance K,

$$A_{phys} = \frac{K}{P_b} \frac{A_{corr} - P_b \Sigma_i f_i A_i}{1 - \Sigma_i f_i} \tag{9}$$

TABLE 1. Corrections to A_{corr} and systematic errors.

Correction (ppb)	Helium	Hydrogen
Beam Asyms.	183 ± 59	-10 ± 17
Target windows bkg.	113 ± 32	7 ± 19
Helium QE bkg.	12 ± 20	-
Rescatter bkg.	0 ± 58	0 ± 15
Scale Factor	**Helium**	**Hydrogen**
Acceptance K	1.000 ± 0.001	0.979 ± 0.002
Q^2 Scale	1.000 ± 0.009	1.000 ± 0.017
Polarization P_b	0.844 ± 0.008	0.871 ± 0.009

Table 1 describes these corrections and summarizes all systematic errors.

HAPPEX RESULTS FOR G_E^S AND G_M^S

The theoretical value of the parity violating asymmetry from helium and hydrogen target, under the assumption that the strange quarks do not contribute, is calculated using citeMusolf:1994, Donnelly:1994. For our kinematics, we obtain :

$$\begin{aligned} A_{NS}^{He} &= +6.37\ ppm && \text{at } Q^2 = 0.007\ GeV^2 \\ A_{NS}^{H} &= -1.66 \pm 0.05\ ppm && \text{at } Q^2 = 0.109\ GeV^2 \end{aligned} \tag{10}$$

Comparing the experimental results to the theretical expectations, the strange form factors are extracted :

$$\begin{aligned} G_E^s &= 0.002 \pm 0.014 \pm 0.007 && \text{at } Q^2 = 0.077\ GeV^2 \\ G_E^s + 0.09 G_M^s &= 0.007 \pm 0.011 \pm 0.004 \pm 0.005_{FF} && \text{at } Q^2 = 0.109\ GeV^2 \end{aligned} \tag{11}$$

where the value $G_E^{(\gamma,n)} = 0.037$ is used from BLAST [14], the contribution from the axial form factor G_A^Z [15] and uncertainties in the nucleon form factors [13].

To illustrate the impact of the Happex measurements on strange form factors at $Q^2 \sim 0.1\ GeV^2$, a small interpolation of each measurement to a common Q^2 was made assuming that $G_E^s \propto Q^2$ and G_M^s remained constant. Figure 1 displays the 2004 and 2005 measurements. The 95% and 68% confidence level ellipses provide a means for comparaison to the theoretical expectations. From the best-fit, the following strange electric and magnetic form factors are obtained :

$$\begin{aligned} G_E^s &= -0.005 \pm 0.019 \\ G_M^s &= 0.18 \pm 0.27 \end{aligned} \tag{12}$$

In summary, we have reported on the extraction of the strange electric and magnetic form factors at $Q^2 \sim 0.1\ GeV^2$ from the measurement of the parity-violating asymmetry

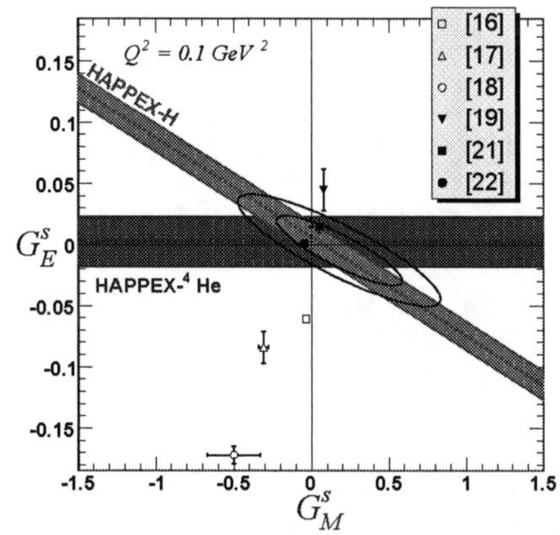

FIGURE 1. Combined data from [10, 11] for G_E^s versus G_M^s at $Q^2 = 0.1\ GeV^2$. All experimental 1-σ bands are the quadrature sum of statistical and systematic errors. Inner and outer ellipses show 95% and 68% confidence level constraints. Various theoretical predictions are plotted [16, 17, 18, 19, 20, 22].

from the elastic scattering of longitudinally polarized electrons from helium and hydrogen targets. These constraints on the strange electric and magnetic form factors at low Q^2 have increased our knowledge and understanding of the nucleon structure.

REFERENCES

1. D. Kaplan and A.V. Manohar, Nucl. Phys. **B310**, 527 (1988).
2. R.D. McKeown, Phys. Lett. **B219**, 140 (1989).
3. D.H. Beck, Phys. Rev. **D39**, 3248 (1989).
4. K.S. Kumar and P.A. Souder, Prog. Part. Nucl. Phys. **45**, S333 (2000).
5. D.H. Beck and B.R. Holstein, Int. J. Mod. Phys. **E10**, 1 (2000).
6. D.H. Beck and R.D. McKeown, Ann. Rev. Nucl. Part. Sci. **51**, 189 (2001).
7. M.J. Musolf et al., Phys. Rep. **239**, 1 (1994).
8. M.J. Musolf, R. Schiavilla and T.W. Donnely, Phys. Rev. **C50**, 2173 (1994).
9. A. Acha, Phys. Rev. Lett. **98**, 032301 (2007).
10. K.A. Aniol et al., Phys. Rev. Lett. **96**, 022003 (2006).
11. K.A. Aniol et al., Phys. Lett. **B635**, 275 (2006).
12. K.A. Aniol et al., Phys. Rev. **C69**, 065501 (2004).
13. J. Friedrich and Th. Walcher, Eur. Phys. J. **A17**, 607 (2003).
14. V. Ziskin, Ph.D. thesis, MIT (2005).
15. S.L. Zhu et al., Phys. Rev. **D62**, 033008 (2000).
16. N.W. Park and H. Weigel, Nucl. Phys. **A541**, 453 (1992).
17. H.-W. Hammer, U.-G. Meissner and D. Drechsel, Phys. Lett. **B367**, 323 (1996).
18. H.-W. Hammer and M.J. Ramsey-Musolf, Phys. Rev. **C60**, 045204 (1999).
19. A. Silva, H.-C. Kim and K. Goeke, Phys. Rev. **D65**, 014016 (2002).
20. R. Lewis, W. Wilcox and R.M. Woloshyn, Phys. Rev. **D67**, 013003 (2003).
21. D.B. Leinweber et al., Phys. Rev. Lett. **94**, 212001 (2005);
22. D.B. Leinweber et al., Phys. Rev. Lett. **97**, 022001 (2006).

The Impact of the BLAST Magnetic Field on the Čerenkov Detectors

B. Tonguc, for the BLAST Collaboration

Department of Physics, Sakarya University, 54187 Adapazari Turkey.

Abstract. We report a factor of two improvement to the efficiencies of a large array of diffusely reflective aerogel Čerenkov detectors which were operated near a high magnetic field. This was accomplished by designing additional magnetic shielding. The improvement was studied by observing the ADC spectrum of the detectors.

Keywords: Čerenkov detectors; Magnetic field; Efficiency; Aerogel.
PACS: 29.40.Ka; 96.12.Hg; 42.60.Lh; 82.33.Ln.

INTRODUCTION

The Bates Large Acceptance Spectrometer Toroid (BLAST) consisted of a number of detector systems. It was located in the South Hall Storage Ring of the MIT-Bates Linear Accelerator Center, Middleton, MA; one of the research laboratories of the Massachusetts Institute of Technology. The facility typically delivered $\sim 70\%$ polarized electrons with energies up to 1 GeV with currents of $\sim$ 175 mA and lifetimes of $\sim$ 25 min. The longitudinal polarization of the beam was maintained by a Siberian Snake and was monitored by a Compton polarimeter. The toroidal magnetic field for BLAST was produced by eight coils and achieved a maximum of $\sim$ 3800 G in the drift region.

The BLAST detector system covered $20-80$ degrees in polar angle and ± 15 degrees in azimuthal angle both left and right of the beamline. Time of flight (TOF) scintillators, Čerenkov detectors, wire chambers, and neutron counters were used for relative timing, particle identification, charged particle tracking, and neutron detection, respectively of the particles produced in the reactions studied (Fig. 1).

The aerogel Čerenkov detectors were used for distinguishing electrons from pions up to momenta of 700 MeV/c. There were four Čerenkov detectors in each sector. The most forward one in each sector had an aerogel radiator 7 cm thick with an index of refraction n = 1.02. The remaining detectors contained 5 cm thick aerogel with n = 1.03. Each detector consisted of a rectangular aluminum box painted with a diffusely reflective material to scatter the Čerenkov light toward the photo-multiplier tubes (PMTs). The PMTs had a 12.7 cm diameter (Fig. 1). The PMTs were shielded against the BLAST magnetic field with two concentric steel cylinders intended to zero the longitudinal magnetic field in the PMT cavity. The inner and outer tubes had 75 mm radius with 10 mm thickness and 88 mm radius with 6 mm thickness, respectively [1].

CP1006, *High Energy Physics and Applications*, edited by J. Ellis, S. Nasri, and E. Malkawi

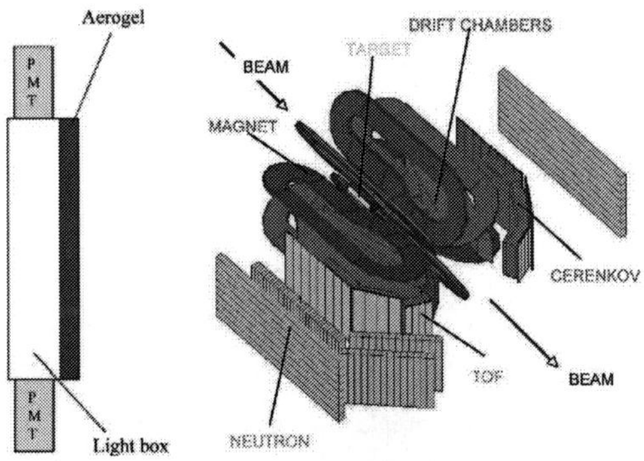

FIGURE 1. A schematic, side view of one of the Čerenkov Counters (left), and an isometric view of the BLAST detector showing the main detector elements (right).

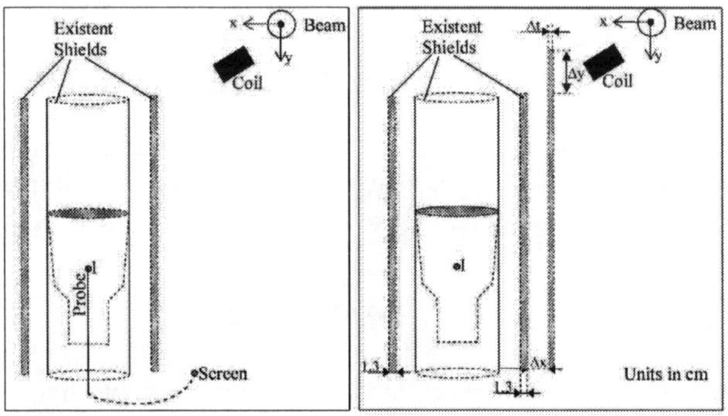

FIGURE 2. Left: cut-away view of the measurement setup. Right: cut-away view of the final shielding design schematic with the iron plate.

RESIDUAL MAGNETIC FIELD, MEASUREMENTS AND IMPLICATIONS

The BLAST Čerenkov detectors were designed to identify electrons with $\sim 90\%$ detection efficiency. However, during data taking, in the presence of the BLAST magnetic field, analyses which compared events from TOF counters to those from Čerenkov detectors yielded $\sim 40-50\%$ efficiencies. Since the PMTs of the Čerenkov counters are very close to the coils of the toroid magnet, the stray magnetic field was thought to be a possible explanation for these low efficiencies.

Magnetic field measurements were carried out inside the PMT cavity with the PMT removed. A F.W. Bell Gauss/Tesla meter, Model 4048 was used. The main purpose of this was to measure the magnetic field inside the PMT cavity to determine if it was large enough to deflect the photoelectrons such that they missed the first dynode in the PMT. A schematic of the measurement setup is shown in Fig. 2.

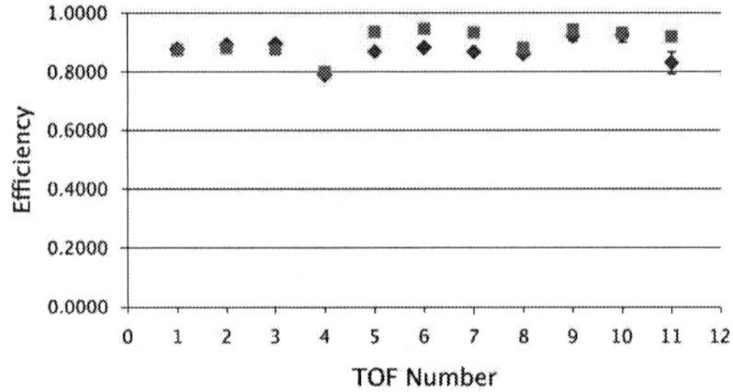

FIGURE 3. Detection efficiency of the Čerenkov counters with respect to the TOF scintillators for both left (rhombus marker) and right (square marker) sectors.

The magnetic field in the PMT cavity was found to be ~ 5 G in the longitudinal direction when the toroidal magnetic field was turned on. Such a value for the magnetic field can reduce the PMT efficiency by up to $\sim 25\%$ [4].

The POISSON SUPERFISH code was used for designing and positioning additional shielding. An additional iron plate between the existing shielding and the magnet coils was chosen for this. Simulation studies showed that the residual magnetic field could be substantially suppressed if the iron plate extended $\Delta y = 10$ cm longer than the PMT shielding tubes, and if it was mounted $\Delta x = 2.5$ cm away (Fig. 2).

DATA COMPARISON

The effect of the new shielding was checked by comparing both the ADC spectra and the efficiency of the Čerenkov detectors for data taken before and after adding the extra shielding. The data consisted mainly of electrons elastically scattered from protons in the hydrogen target and were selected based on coplanarity and timing cuts (Fig. 3). A substantial difference could be seen between the ADCs before and after the shielding (Fig. 4). The "after" spectra had wider peaks with rather short tails. The ADC spectra were fit with a Poisson function,

$$y = (Const) \times \frac{\mu^x e^{-\mu}}{\Gamma[x+1]}, \tag{1}$$

where the fit parameter μ corresponds to the average photon yield per event [3]. In Eq. 1 y, $Const$, x, and Γ are: the number of counts per channel, a normalization factor, the ADC channel number, and the gamma function, respectively.

SUMMARY AND CONCLUSIONS

Initial operation of the BLAST aerogel Čerenkov detectors at the MIT-Bates Linear Accelerator Center were not the same as had been observed earlier during testing when the only source of magnetic field was the earth. Physical magnetic field measurements

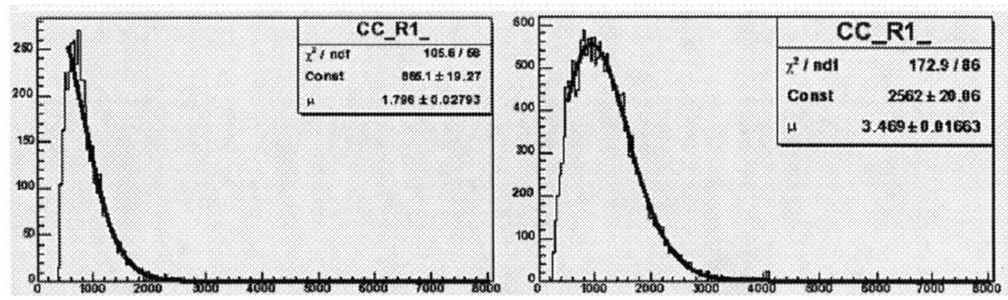

FIGURE 4. The ADC spectrum of Čerenkov box CC_R1 and Poisson function fit (solid line) for before (left) and after (right) the additional shielding.

in the PMT region of the detector, when the BLAST magnetic field was on, revealed the presence of a nontrivial longitudinal field bypassing the existing shielding. Additional shielding in the form of a simple iron plate was practical. The effect of the new shielding was checked by comparing the magnetic field measurements and ADC spectra of the detector before and after shielding.

It was observed that after the additional shielding was installed, the longitudinal magnetic field was reduced to less than 0.5 gauss corresponding to $\sim 5\%$ inefficiency for the PMTs. In addition, after the shielding, the ADC spectrum of the Čerenkov detector box (CC_R1), the middle one on the right, gave typically an average of 3.4 photoelectrons per event corresponding to $\sim 95\%$ detection efficiency [2]. This result agreed well with the efficiency result obtained by comparing the counts of electrons for both the Čerenkov box and the TOF scintillators. Similar improvements were observed for the other Čerenkov boxes.

REFERENCES

1. B. Tonguc, et al., *Nucl. Instr. And Meth.* **A**, no. 553, 364–369 (2005).
2. D.W. Higinbotham, *Nucl. Instr. And Meth.* **A**, no. 414, 332 (1998).
3. L.C. Alexa et al., *Nucl. Instr. And Meth.* **A**, no. 365, 299 (1995).
4. anon, "Operating Characteristics and Considerations" in *Photomultiplier Tubes: Principles and Applications*, Philips Photonics, Brive, 1994, pp. 4.51–4.52.

Status of ATLAS and CMS

Albert De Roeck

CERN, 1211 Geneva 23, Switzerland and Antwerp University, Belgium

Abstract. The status of the CMS and ATLAs experiments is given, as it is end of 2007. A quick tour of possible new physics is presented.

Keywords: LHC, ATLAS, CMS, Higgs, Beyond the Standard Model
PACS: 12.60.-i,12.38.-t,13.85.-t

INTRODUCTION

The Large Hadron Collider LHC will be a unique tool for fundamental physics research and the highest energy accelerator in the world for many years following its completion. The LHC will provide proton-proton collisions at $\sqrt{s} = 14$ TeV. The event rate is about 1,000 times higher than that presently achieved at the Tevatron $p\bar{p}$ collider. The first collisions at 14 TeV are expected for early summer 2008.

The physics potential of the LHC is unprecedented: it will allow to study directly and in detail the TeV scale region. The LHC is expected to shed light on the electroweak symmetry breaking mechanism (EWSB) and provide evidence of physics beyond the Standard Model (SM) [1]. The LHC will be also a Standard Model precision measurement instrument, mainly due to the very high event rates as shown in Table 1.

TABLE 1. Approximate event rates for some physics processes at the LHC for a luminosity of $L = 2 \times 10^{33}\,\mathrm{cm}^{-2}\mathrm{s}^{-1}$. For this table, one year is equivalent to 20 fb^{-1}.

Process	Events/s	Events/y
$W \to e\nu$	40	$4 \cdot 10^8$
$Z \to ee$	4	$4 \cdot 10^7$
$t\bar{t}$	1.6	$1.6 \cdot 10^7$
$b\bar{b}$	10^6	10^{13}
$\tilde{g}\tilde{g}$ (m = 1 TeV)	0.002	10^4
Higgs (m= 120 GeV)	0.08	$8 \cdot 10^5$
Higgs (m= 800 GeV)	0.0012	$1.2 \cdot 10^4$
QCD jets $p_T >$ 200 GeV	10^2	10^9

The proton beams cross at interaction points along the ring where detectors that measure the particles produced in the collisions are installed. Interaction point 5 hosts the CMS detector, shown in Fig. 1. Interaction point 1 is the cavern of the ATLAS experiment shown in Fig. 2. ATLAS and CMS are general multi-purpose detectors, with the mission to discover, or exclude within the SM, the Higgs particle in the full range of interest, and thus shed light on the mechanism of electroweak symmetry breaking

CP1006, *High Energy Physics and Applications*, edited by J. Ellis, S. Nasri, and E. Malkawi

[2, 3]. Furthermore, the LHC will be the first machine that allows to study the Tera-energy scale, and has excellent chances to discover physics beyond the SM. The broad capabilities of CMS and ATLAS are tailored for the detection of these new phenomena and particles.

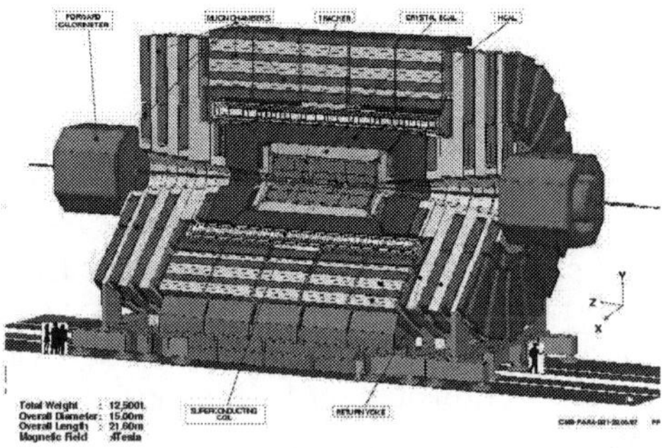

FIGURE 1. Three dimensional view of the CMS detector, and its detector components.

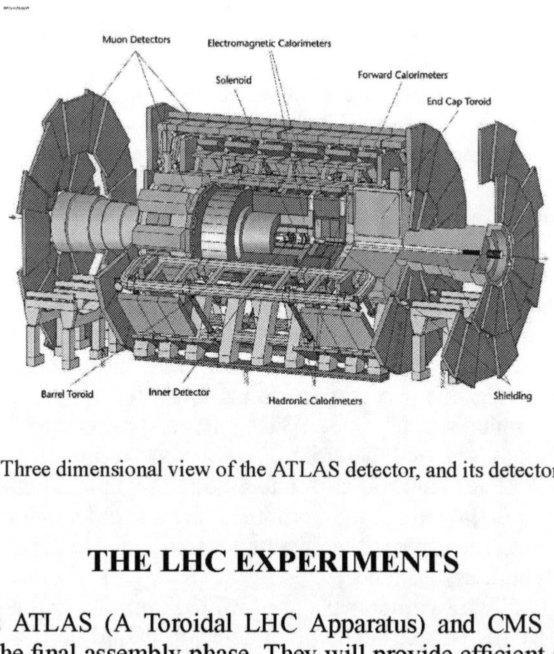

FIGURE 2. Three dimensional view of the ATLAS detector, and its detector components.

THE LHC EXPERIMENTS

Two experiments ATLAS (A Toroidal LHC Apparatus) and CMS (Compact Muon Solenoid) are in the final assembly phase. They will provide efficient and precise measurements of electrons, muons, taus, neutrinos, photons, light flavour jets, b-jets and the missing transverse momentum. A brief summary of the main design choices is given in the following. The main features of the two experiments are summarized in Table 3 at the end.

CMS is centered around one magnet, a big solenoid, which contains the inner detector and the calorimeters and provides a magnetic field of 4 T in the inner detector volume. ATLAS has four magnets: a solenoid sitting in front of the electromagnetic calorimeter and producing a field of 2 T in the inner cavity, and external barrel and end-cap air-core toroids. The general structure of the two detectors is determined by this basic design choice.

Both detectors reconstruct the tracks from the interaction vertex using inner detectors built of layers of silicon pixel and silicon strip detectors immersed in the solenoidal magnetic field. These detectors provide a high precision measurement of the vertex position and of the track momenta. While the CMS tracking detector is all silicon, the ATLAS inner detector in addition contains a Transition Radiation Detector (TRT) at larger radii. Excellent momentum resolution is expected for both experiments as shown in Table 3. Due to the lower magnetic field the expected momentum resolution in ATLAS is a factor of about two worse than that of CMS. However, the Transition Radiation Detector provides electron/pion separation capabilities. Thanks to silicon strip and pixel detectors, both experiments will be able to perform b quark tagging using impact parameter measurements and the reconstruction of secondary vertices.

The CMS electromagnetic calorimeter is a high-resolution lead tungstate ($PbWO_4$) crystal detector. The ATLAS calorimeter is a lead-liquid argon sampling calorimeter, therefore with a worse intrinsic energy resolution. However, thanks to a very fine lateral and good longitudinal segmentation, the ATLAS calorimeter potentially has more robust particle identification capabilities.

In both experiments the hadronic calorimeters are sampling detectors with scintillator or liquid-argon as active medium. The ATLAS hadronic calorimeter offers a better energy resolution because it is thicker (the CMS hadronic calorimeter is constrained in space as dictated by the external solenoid dimensions) and has a finer sampling. CMS will be using particle flow techniques to improve the resolution of the calorimeter response. For both experiments the calorimetric coverage is extended down to a rapidity of $|\eta| < 4.9$ in order to ensure hermetic coverage for the measurement of missing transverse momentum, and to allow the detection of forward jets.

Finally, the external Muon spectrometer of CMS consists of chamber stations embedded into the saturated iron of the solenoid return yoke, thus achieving a design where a single magnet provides the necessary bending power for precise tracking in the inner detector and in the muon spectrometer. ATLAS has a spectrometer in air, where multiple scattering is minimised, and therefore offers the possibility of good standalone (i.e. without the inner detector contribution) muon momentum measurement with three stations of high-precision tracking chambers at the inner and outer radius and in the middle of the air-core toroid. The expected momentum resolution is better than 10% for muons of p_T= 1 TeV/c in both experiments.

For CMS essentially all components are lowered and installed except for the pixel detector and the ECAL endcaps. The pixel detector should be installed and be ready for day one, but it is still open if both ECAL endcaps will be ready in time for the first collisions. For ATLAS the last big component, a small wheel, was lowered in the cavern on February 29 2008. The wall wheels of the muon detector are still being assembled. Both experiments are now in a frantic campaign to commission the detector and get it as operational as possible before the first collisions come. Cosmics are a welcome tool for such studies, and beam halo or beamgas collisions wil take that role by middle of 2008

The challenges at the LHC are enormous. The total inelastic cross section at the LHC is expected to be close to 80 mb. The LHC will operate at a bunch crossing rate of 40 MHz, but only 80 % of the bunches will be filled. The instantaneous luminosity in the first two years after start–up is expected to be $L = 2 \times 10^{33}$ cm^{-2}s^{-1} and subsequently upgraded to $L = 10^{34}$ cm^{-2}s^{-1} in a second phase. The average number of inelastic non-diffractive interactions per bunch crossing are respectively 4 and 20 events. These form the so called pile-up interactions.

A complex trigger system is needed to reduce the $10^8 - 10^9$ interaction rate, to a data recording rate of 100-200 Hz. CMS and ATLAS have $\sim 10^8$ channels that are checked each bunch crossing. The design data-size per event is about 1 MB. At start-up it is essential to allow for a larger event size, up to 1.5 MB per event, in order to be able to thoroughly study and understand the detector performance.

At the LHC more than 10 PetaBytes of data will be produced per year. These data need to be stored, processed and be accessible from all over the world. LHC experiments count on the computing GRID to play that role. The Grid will consist of more than 100,000 processors, multi-PentaByte disk and tape storage. The Grid is based on a Tier Architecture (see [4]).

FIRST PHYSICS

Early measurements

According to the schedule of the machine, the first delivered luminosity will be of order of 10^{30}cm^{-2}s^{-1}, i.e. much lower than the values mentioned before. Table 1 shows clearly that even with, say 10 pb^{-1} of integrated luminosity, there will be already a large number of di-jet, W,Z and top quark events. The first events the LHC experiments will detect are soft hadronic events, $pp \to X$ with X a final state of hadrons with an average p_T of a few hundered MeV/c, up to tails of a few GeV/c.

Our knowledge on what to expect at the LHC for soft hadronic events is limited and based on phenomenological models and Monte Carlo programs. In fact the predictions for the charged particle density in pp collisions at 14 TeV differ by 20-30% between different models or assumptions, see Fig. 3. Hence probably one of the very first physics papers at the LHC will be the measurement of the charged particle density in soft pp collisions. These are studies that can be performed with the smallest of luminosities, but of course need a reasonably well understood tracker.

Next we will see the first hard scattering events, in form of jets and di-jet events. Jet cross section measurements are a must for each new collider. The high E_T part of the jet spectrum could contain signals for new physics. Even with 10 pb^{-1} we expect already O(100) di-jet events with a jet $E_T > 1$ TeV. With time and accumulated luminosity many QCD topics will be studied, such as the aforementioned jet studies, extraction of parton density information, jet shape studies, diffractive studies, BFKL studies and perhaps even the extraction of α_s.

From Table 1 it can be seen that already with 10 pb^{-1} we expect of the order of $10^6 W$, $10^5 Z$ and 10^4 top quark pairs to be produced. Before the LHC experiments can launch

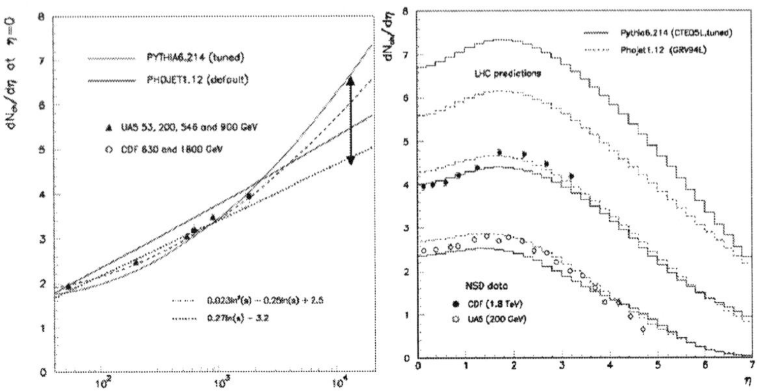

FIGURE 3. Expected charged particle density (left) and rapidity distribution of charged particles (right) as predicted by several models for pp collisions at 14 TeV. The precision of pseudo-data is shown as well.

into a dedicated search for new physics they will first have to *rediscover the Standard Model*, by measuring the cross sections of these particles and more. Next will be the measurement of the more complicated topologies, such as tt+njets, W+njets, Z+njets. These topologies are important background channels for e.g. SUSY searches and need therefore to be well measured and characterised.

Searching the Higgs particle

One of the key questions in particle physics is the origin of electroweak symmetry breaking, eg why is the photon massless while the Z is very massive? The most elegant explanation within the SM is a Higgs field with at least one scalar particle, the Higgs boson. The LHC search reach has been largely optimized for finding the SM Higgs particle, or excluding its existence. Table 2 shows the possible discovery channels for the Higgs at the LHC in the low mass range ($M_H < 200$ GeV/c^2). In the intermediate and high mass range in particular the channels $H \rightarrow WW, H \rightarrow ZZ$ are important with leptonic decays, but at high mass also with jet decays of the vector bosons.

The experimental reach of the the CMS experiment at the LHC is shown in Fig. 4 (left) for the most significant channels. Reversely Fig. 4 (right) shows what luminosity is needed to exclude with combined CMS and ATLAS data the Higgs hypothesis as function of mass. Clearly the first 1 fb^{-1} will already be very revealing.

Hence the Higgs program at the LHC looks as follows. The SM Higgs will be discovered in the full region up to 1 TeV or its existence will be excluded with O(10) fb^{-1} or less. If no Higgs is observed, other new phenomena in the WW scattering should be observed around 1 TeV. The LHC will measure with full luminosity (100-300 fb^{-1}) the Higgs mass with 0.1-1% precision; the Higgs width, for $m_H > 200$ GeV/c^2 with 5-8% precision; the Higgs cross sections times branching ratios with 5-20% precision; ratios of couplings with 10-30% precision absolute couplings only with additional assumptions; spin information in the ZZ channel for $m_H > 200$ GeV/c^2; CP

TABLE 2. Production and decay modes for Higgs particles with mass less than 200 GeV that have been studied at the LHC

	Production			
Decay	Inclusive	VBF	WH/ZH	ttH
$H \to \gamma\gamma$	yes	yes	yes	yes
$H \to bb$			yes	yes
$H \to \tau\tau$		yes		
$H \to WW^{(*)}$	yes	yes	yes	
$H \to ZZ, Z \to ll$	yes			
$H \to Z\gamma, Z \to ll$	low σ			

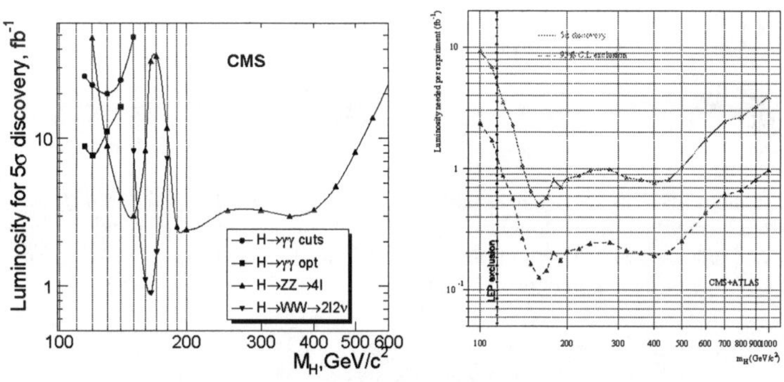

FIGURE 4. (Left)The integrated luminosity needed for the 5σ discovery of the inclusive Higgs boson production in the most promising channels; (Right) The prospects for discovering a Standard Model Higgs boson in initial LHC running, as a function of its mass, combining the capabilities of ATLAS and CMS.

information from exclusive central production $pp \to pHp$.

In general we will get a pretty good picture of the Higgs at the LHC. Even more detailed information can be extracted from a high energy e^+e^- collider.

Beyond the Standard Model

The second most important task of the LHC is the search for new physics beyond the Standard Model. New physics is expected –but fully not guaranteed – around the TeV scale. It can provide answers to questions such as stabelizing the Higgs mass, the hierarchy problem, unification gauge couplings, dark matter... Two popular extensions of the Standard Model are supersymmetry and extra dimensions. However there is whole plethora of possibilities eg. Little Higgs models, split supersymmetry, new gauge bosons, technicolor, compositness, leptoquarks, unparticles, valley physics, etc. All these possible scenarios, will leave measurable traces in collisions at the LHC.

Will new discoveries show up easily a the LHC? Fig. 5 shows a di-lepton resonance at a mass of 1 TeV/c^2 showing up in the di-lepton spectrum. The mere fact that it sticks out as a peak and not just a global enhancement of the background is extremely helpful for a fast discovery. If this happens, LHC could be lucky and already see signals of new physics very early on. Also e.g low mass SUSY could show up fast in the data, but it may take some time before we are confident that it is actually SUSY what has been detected.

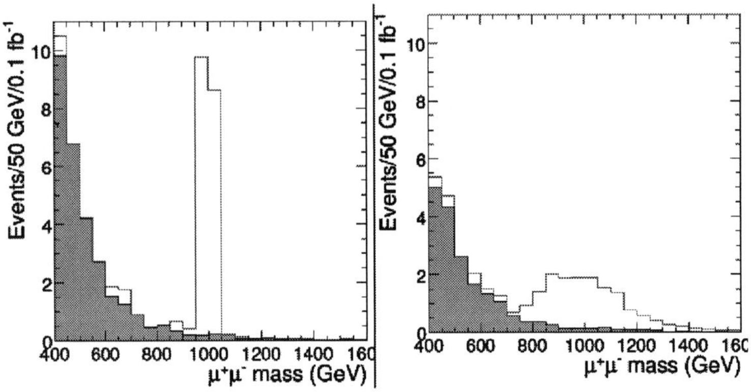

FIGURE 5. Histograms of the $\mu^+\mu^-$ invariant mass for 1 TeV/c^2 Z' plus background (open histogram) and for background only (shaded histogram), at the event-generator level (left) and for events selected by the triggers and reconstructed assuming the "first data" misalignment scenario (right). The number of events per bin is normalised to an integrated luminosity of 0.1fb^{-1}.

SUMMARY

In the current schedule the CMS and ATLAS detectors will be largely ready for the first collisions in summer 2008. The first physics at the LHC promises to be very interesting. After a dedicated period of detector commissioning the first Standard Model "rediscovery" measurements will be made, in terms of jets, vector bosons and top quarks. The hunt for finding the Higgs will be on but the potential to discover it at an early stage depends strongly on its mass. New physics signatures could also show up very early. Will this be the case at the LHC? In 2008/09 we will finally know!

REFERENCES

1. J.G. Branson et al., *High Transverse Momentum Physics at the Large Hadron Collider*, Eur. Phys. J Direct **C4** (2002) N1, hep-ph/0110021.
2. ATLAS Collaboration, *ATLAS detector and physics performance Technical Design Report*, CERN/LHCC 99-14/15 (1999). http://atlas.web.cern.ch/Atlas/GROUPS/PHYSICS/TDR/access.html.
3. CMS Collaboration *CMS physics : Technical Design Report* Volume 2 : Physics performance, CMS TDR 8.2 CERN-LHCC-2006-021 (2006); J. Phys. G: Nucl. Part. Phys. 34 (2007) 995, doi:10.1088/0954-3899/34/6/S01
4. I. Bird, this meeting.

TABLE 3. Main features of the ATLAS and CMS detectors.

	ATLAS	CMS
Magnet(s)	Air-core toroids + solenoid in inner cavity	Solenoid
	Calorimeters in field-free region	Calorimeters inside field
	4 magnets	1 magnet
Inner detector	Si pixels and strips	Si pixels and strips
	TRT $\rightarrow$ particle identification	No particle identification
	B = 2 T	B = 4 T
	$\sigma/p_T \sim 3.4 \times 10^{-4} p_T(\mathrm{GeV}) \oplus 0.01$	$\sigma/p_T \sim 1.5 \times 10^{-4} p_T(\mathrm{GeV}) \oplus 0.008$
EM calorimeter	Lead-liquid argon	$PbWO_4$ crystals
	$\sigma/E \sim 10\%/\sqrt{E(\mathrm{GeV})}$	$\sigma/E \sim 3-5\%/\sqrt{E(\mathrm{GeV})}$
	Longitudinal segmentation	No longitudinal segmentation
HAD calorimeter	Fe-scintillator + Cu-liquid argon	Brass-scintillator
	$\geq 10\ \lambda$	$\geq 7.2\ \lambda$ + tail catcher
	$\sigma/E \sim 50\%/\sqrt{E(\mathrm{GeV})} \oplus 0.03$	$\sigma/E \sim 100\%/\sqrt{E(\mathrm{GeV})} \oplus 0.05$
Muon spectrometer	Chambers in air	Chambers in solenoid return yoke (Fe)
	$\sigma/p_T \sim 7\%$ at 1 TeV	$\sigma/p_T \sim 5\%$ at 1 TeV
	spectrometer alone	combining spectrometer and inner detector

Experimental Tests of Discrete Symmetries with Kaons and B-mesons

Bernard Peyaud

CEA-Saclay, DAPNIA, Gif sur Yvette 91191 Cedex France

Abstract. During the last decades, experiments using K mesons have provided precise measurements of CP violation parameters which are important inputs for the understanding of flavour physics. They include the measurements of Re(ε'/ε) by NA48 and KTeV experiments using K_L and K_S, the search for CP violation in $K^{\pm} \rightarrow 3\pi$ decays and rare decay processes allowing to search for new physics and to test chiral perturbation theory. Moreover, NA62, a new experiment at the CERN/SPS, will measure the branching ratio of $K^+ \rightarrow \pi^+ \nu \bar{\nu}$ to test further the Standard Model. Experimental efforts in the kaon sector will thus continue for some time. Alternatively, in the B-meson system, there are more decay modes available, and the Standard Model makes precise predictions for CP violation in a number of these. The B-meson system is therefore a very attractive area to study CP violation and to search for new physics. The LHCb experiment at the Large Hadron Collider (LHC) and its prospects in the study of CP violation in the B-meson sector are discussed.

Keywords: CP violation, K and B mesons
PACS: 11.30.Er, 13.25Es, 13.25Hw, 14.40Aq

INTRODUCTION

The discovery of CP violation was made in the neutral kaon system which continues to be a great resource to study this fundamental phenomenon of Nature. It was quickly established that the dominant part of the process comes from *indirect* CP violation where the physical states K_L and K_S deviate slightly from the K_2 (CP=-1) and K_1 (CP=1) eigenstates. This mixing is characterized by the complex parameter $\varepsilon \approx 2.3.10^{-3}.\exp^{i45^\circ}$ which fixes the size of the deviation from pure CP states. Within the Standard Model (SM) ε is related to the phase δ of the CKM mixing matrix describing the 3 generations of quarks. There is also a possibility for *direct* CP violation in $K_2 \rightarrow \pi\pi$ decays and it is fundamental to understand this possible effect within the SM. Direct CP violation in the kaon system is parameterized by ε' and corresponds to different decay amplitudes for K_0 and $\bar{K}_0$. When expressed to first order in terms of ε and ε' the amplitude ratios for neutral decays $\eta_{00} \equiv A(K_L \rightarrow \pi^0\pi^0)/A(K_S \rightarrow \pi^0\pi^0)$ and for charged decays $\eta_{+-} \equiv A(K_L \rightarrow \pi^+\pi^-)/A(K_S \rightarrow \pi^+\pi^-)$ are given respectively by $\eta_{00} = \varepsilon - 2\varepsilon'$ and by $\eta_{+-} = \varepsilon + \varepsilon'$. The measurement of the double ratio $R = |\eta_{00}/\eta_{+-}|^2$ gives access to direct CP violation through R=1-6 Re(ε'/ε) which has received a lot of theoretical and experimental investigations.

CP1006, *High Energy Physics and Applications*, edited by J. Ellis, S. Nasri, and E. Malkawi

THE DISCOVERY OF CP VIOLATION

In 1964, a remarkable experiment at Brookhaven National Laboratory (BNL) reported evidence of the long-lived neutral kaon decaying in the forbidden state $\pi^+\pi^-$. Under the assumption of CP symmetry the model [1] based on K_0 mixing with $\bar{K}_0$ predicts the long lifetime state K_2 with eigenvalue CP=-1 and therefore is not allowed to decay to 2π which has CP=+1. In 1956, the first observation [2] of K_2 was made at BNL in a bubble chamber. The typical observation was the allowed mode $K_2 \rightarrow \pi^+\pi^-\pi^0$ and also the semileptonic modes K_{e3} and K_{l3}. None of 186 observed events could be kinematically fitted under the assumption of $K_2 \rightarrow \pi^+\pi^-$ decay and, as expected, CP was conserved to high accuracy. In 1961, a similar bubble chamber experiment [3] conducted at JINR-Dubna reported on 225 decays with no sign of $K_2 \rightarrow \pi^+\pi^-$ decay and the conclusion was reached that the branching ratio could not exceed 1/400. Then, in 1963, a liquid hydrogen bubble chamber was placed in a neutral beam at the BNL Cosmotron and 252 K_2 decays were observed. These events were analyzed as $K_2 \rightarrow \pi^+\pi^-$ decays through their angular distribution where a forward peak of 19 events was given the interpretation of "anomalous regeneration" [4]. This perplexing result was one of the reasons that led the Princeton team in April 1963 to submit a Letter of Intent for an experiment based on an existing apparatus, a double-armed spectrometer with spark chambers and Cerenkov counters as shown in Figure 1. The BNL Directorate gave agreement in May 1963 and the apparatus was installed and ready on June 2nd. The run lasted until the end of July with 70000 triggers taken for regeneration studies, 47000 triggers for CP limit and 23000 triggers to study the "Adair effect" on hydrogen. By looking at mass slices below, above and at the K_0 mass a peak in the forward direction was detected during a first analysis made between October and December 1963. The team decided to reanalyze the events with more accuracy in the measurement of the spark positions. From January to April 1964 many attempts were made to "kill" the peak but none was successful. One year after they had submitted their proposal the Princeton group demonstrated to the scientific community that the $\pi^+\pi^-$decay of the long lived kaon was occurring with a branching ratio of $(2\pm0.4)\mathrm{x}10^{-3}$. This unexpected discovery [5] was "made with a simple apparatus, designed to ask the right question of nature and received a clear response".

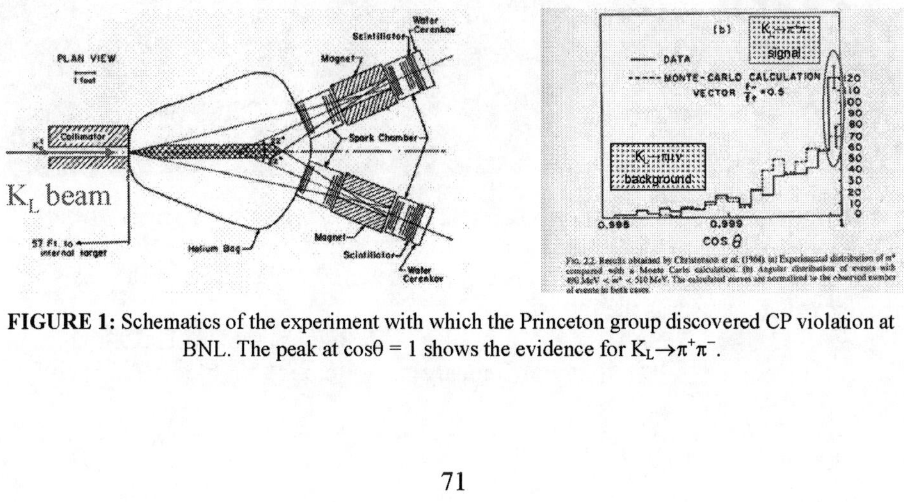

FIGURE 1: Schematics of the experiment with which the Princeton group discovered CP violation at BNL. The peak at cosθ = 1 shows the evidence for $K_L \rightarrow \pi^+\pi^-$.

THE QUEST FOR DIRECT CP VIOLATION

Super weak theory is wounded by NA31

If ε is the only source of CP violation we would expect $\eta_{00}=\eta_{+-}$.as predicted by the super-weak hypothesis [6]. In 1979 a proposal was made at Fermilab to measure the ratio R of CP violating rates $|\eta_{00}|^2$ and $|\eta_{+-}|^2$ to a precision better that 1%. The E617 experiment achieved 3% accuracy only but paved the way for the next measurement made by E731. The principle of the experiment relies on the use of two neutral beams with one of them incident on a regenerator (Figure 2a) thus allowing simultaneous study of K_S/K_L decay products in the apparatus. The most important parts of the detector are drift chambers to measure the momenta of charged particles, lead-glass electromagnetic calorimeter to measure γ's from π^0 and muon hodoscope. Another proposal was made in 1980 at CERN with a different principle where neutral and charged rates are measured simultaneously in the NA31 detector which consist of proportional chambers to measure the charged directions, liquid argon photon calorimeter, hadron calorimeter and muon veto. The simultaneous K_S and K_L beams are obtained by using a movable target (Figure 2a) inside a long vacuum tunnel. Both detectors use high performance electromagnetic calorimeters (i.e. good γ position and energy resolution) in order to reject the main background from $3\pi^0$ decays in the neutral mode. In the charged mode E731 achieves good background rejection of K_{e3} by combining momentum measurement in the drift chamber spectrometer and energy measurement in the electro-magnetic calorimeter (E/p identification of electron) whereas NA31 obtains good π identification with the hadronic calorimeter.

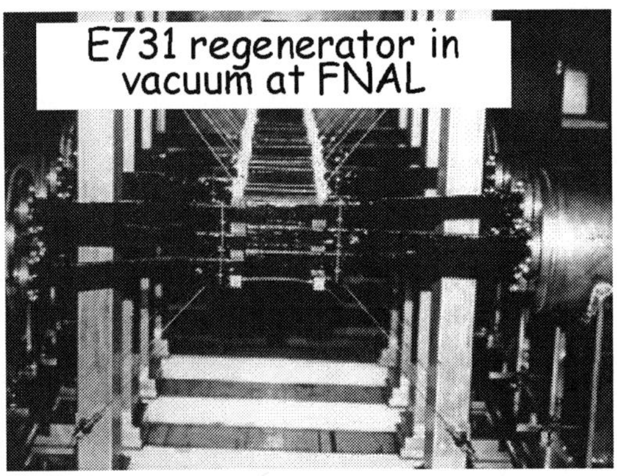

FIGURE 2: a) The E71 carbon regenerator used at FNAL and b) the NA31 K_S target at CERN.

At the concluding talk of the Blois Conference in 1989 the status of the analysis of the two experiments was the following:

NA31	$Re(\varepsilon'/\varepsilon) = (33\pm11)x10^{-4}$,
E731 (very preliminary)	$Re(\varepsilon'/\varepsilon) = (-10 \text{ to } 10 \pm15)x10^{-4}$,

The evidence of a non-zero effect was to some extent diluted by the variance between the results of NA31 and E731 and it was concluded that the experiments must be continued to still higher accuracy.

The magnitude of direct CP violation signal was still contentious when the analysis of the two experiments [7] , [8] were completed in 1996 with the final results:

NA31 $\quad Re(\varepsilon'/\varepsilon) = (23\pm6.5)x10^{-4}$,

E731 $\quad Re(\varepsilon'/\varepsilon) = (7.4\pm6.5)x10^{-4}$.

The next round of experiments on $Re(\varepsilon'/\varepsilon)$ had already started in the early 90's at CERN with NA48 and at FNAL with KTeV.

The experimental fix of ε'/ε

The NA48 and KTeV experiments were designed to measure $Re(\varepsilon'/\varepsilon)$ with an accuracy of $\leq 2x10^{-4}$. In order to reach this requirement the limiting mode $K_L \rightarrow \pi^0\pi^0$ must contain at least 4 million events and the systematic errors be kept at very small levels.

The principle of the NA48 experiment is to use almost collinear K_L and K_S beams pointing at the centre of the detector and having similar momentum spectra ($<P_K>$=110GeV/c) and to detect all 4 decay modes at the same time. Large statistics require high-intensity beams and therefore high-rate capability and performance for the detector elements. The identification of the neutral/charged mode is made with high-resolution photon-calorimeter/spectrometer and K_L and K_S are separated by tagging. By applying a lifetime weighting procedure to the K_L decays in both modes one reduces almost completely the acceptance differences with K_S due to their lifetimes. The weighted vertex distributions of K_L are made similar to the K_S ones thus bringing the experiment close to "gedanken" situation with K_L/K_S coming from the same source.

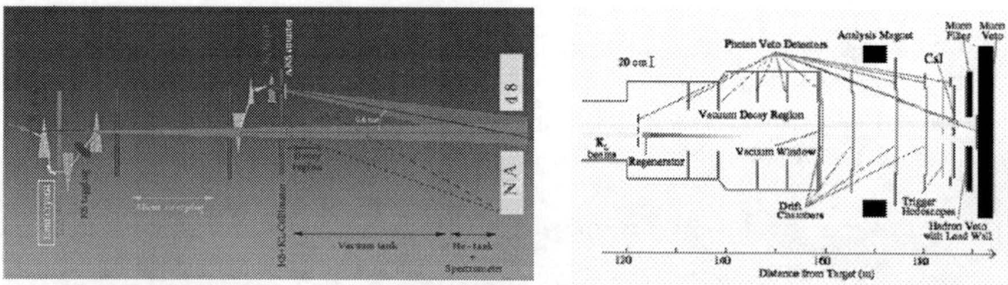

FIGURE 3: a) the K_L/K_S beam line used at CERN by NA48
b) the K_L and regenerated K_S used at FNAL by KTeV.

A schematic view of the neutral kaon beam lines and of the detector layout is shown in Figure 3a. The decay region extends over 90m inside a vacuum tank terminated by a thin Kevlar window. It is followed by the detector elements installed between 216m to 255m downstream of the K_L target. A magnetic spectrometer is inside a helium tank and is followed by a hodoscope to trigger on the charged mode, a liquid krypton (LKr) calorimeter, a hadronic calorimeter, and a muon detector.

The principle of the KTeV experiment is to use a double K_L beam and to produce K_S by regeneration of one of them as shown in Figure 3b. The state of the art detector is a photon calorimeter made of pure CsI crystals. The decay region is shorter here owing to the fact that kaons are less energetic ($\langle P_K \rangle = 70$GeV/c). For both detectors the mass resolution for $K_{L,S} \to \pi^+\pi^-$ decays is ~2MeV/c^2 and the time resolution of the hodoscopes is ~200ps.

Direct CP violation has been measured [9] , [10] by both experiments (Figure 4a) and Re(ε'/ε) is now known to be $(16.3\pm2.2).10^{-4}$. This non-zero value established with 7 standard deviations provides a second number to describe CP violation in the neutral kaon system. It translates in the matter antimatter asymmetry ratio:

$$\frac{\Gamma(K^0 \to \pi^+\pi^-) - \Gamma(\overline{K}^0 \to \pi^+\pi^-)}{\Gamma(K^0 \to \pi^+\pi^-) + \Gamma(\overline{K}^0 \to \pi^+\pi^-)} = (5.04 \pm 0.82) \times 10^{-6}$$

These successes have been made possible by a remarkable sequence of ambitious experimental programs in Europe and USA as summarized in Table1.

TABLE 1. Data statistics of experiments at FNAL and CERN

Year	experiment	$K_L \to 2\pi$	Background	Data volume
1985	E617 FNAL	3,152	~11.3%	10GB
1988	E731 test FNAL	6,950	~5%	
1988	NA31 CERN	109,000	~4%	
1990	E731A FNAL	52,000	~5%	
1993	E731 Final FNAL	410,326	~5%	100GB
1993	NA31 Final CERN	428,000	~2.7%	
2001	KTeV FNAL	10,000,000	~0.48%	
2001	NA48 CERN	5,000,000	~.16%	100TB

On the theoretical side ε'/ε is very hard to calculate because the sensitivity is in the sizes and phases of similar amplitudes [11] There is agreement that direct CP violation effect is much smaller than the indirect one. Despite huge efforts, ε'/ε is not yet computed reliably but improvements are expected from lattice calculation

Charged kaons: another attack by NA48

While at very small level, direct CP violation is also expected for charged kaon decays in the standard model, and NA48 has used intense $K^\pm$ beams in 2003/04 to measure the Dalitz plot properties of the $\pi^\pm\pi^+\pi^-$ and $\pi^\pm\pi^0\pi^0$ decays modes. If CP invariance holds strictly the matrix elements $|M(u,v)|^2 \propto 1 + g^\pm u + hu^2 + kv^2$ for $K^\pm$ must be identical with $g^+ = g^-$ whereas the SM asymmetry $A_g=(g^+-g^-)/(g^++g^-)$ is expected to be in the range (10^{-5} to 10^{-4}). This precision measurement [12] was done by NA48 with final samples of $3.1\text{x}10^9$ $K^\pm \to \pi^\pm\pi^+\pi^-$ decays and $91\text{x}10^6$ $K^\pm \to \pi^\pm\pi^0\pi^0$ decays giving respectively the following results [13]:

$$A_g(\pi^\pm\pi^+\pi^-) = (-1.5 \pm 2.1)\cdot 10^{-4} \text{ and } A_g(\pi^\pm\pi^0\pi^0) = (-1.5 \pm 2.1)\text{x}10^{-4}$$

Compared to previous experiments these results are 10 times more accurate and the errors are dominated by statistics. No CP violation was found but the window to search for new physics has been reduced dramatically (Figure 4b). J. Cronin's hope of 1980 "that at some epoch, perhaps distant, this cryptic message from Nature will be deciphered" has been significantly but partially satisfied.

Another important exploration of CP violation is being prepared by experiment NA62 at CERN [14]. This is the search for the very rare decay $K^+ \to \pi^+ \nu \bar{\nu}$ which is a CKM suppressed process occurring through Flavour Changing Neutral Current. It is very clean theoretically since short distance contributions are dominant and hadronic matrix elements are obtained from the measured decay $K^+ \to \pi^0 l \nu$. The SM prediction of the branching ratio is $BR(K^+ \to \pi^+ \nu \bar{\nu}) = (8.0 \pm 1.1) \times 10^{-11}$ for which the experimental measurement made by E787/949 is $BR(K^+ \to \pi^+ \nu \bar{\nu}) = (1.47^{+1.30}_{-0.89}) \times 10^{-10}$ (3 events).

The NA62 experiment aims at observing O(100) $K^+ \to \pi^+ \nu \bar{\nu}$ events with a background level not exceeding 10%. The sensitivity to new physics is again the driving force of this enterprise.

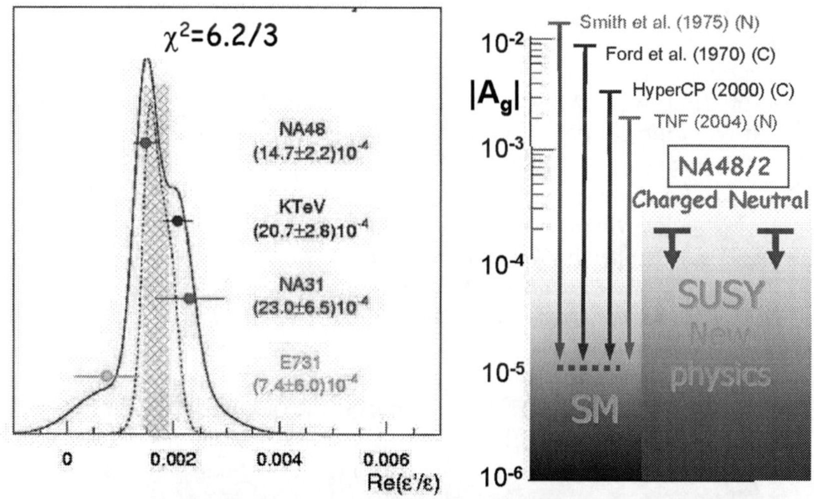

FIGURE 4: a) Direct CP violation measurements of Re(ε'/ε)
b) Limits on Ag from $K^{\pm}$ asymmetry measurements.

THE USE OF B MESONS

Neutral kaon mixing has played a major role in understanding how the CP violating phase δ enters the SM. Similarly Bd mixing was the first experimental phenomenon to correctly anticipate the large value of the top quark mass and later the precise measurements of many different B-meson decay modes obtained in the B-factories, allowed the CKM picture of mixing and CP violation to be verified at the few per cent level. The unitarity triangle (Figure 5a) is known today with good accuracy from the measurement of |Vcb/ Vub|, $\Delta m(B_d)$, and the observation on $\Delta m(B_s)$. While these quantities are CP-conserving, they imply non-zero angles of the triangle, and therefore CP violation.

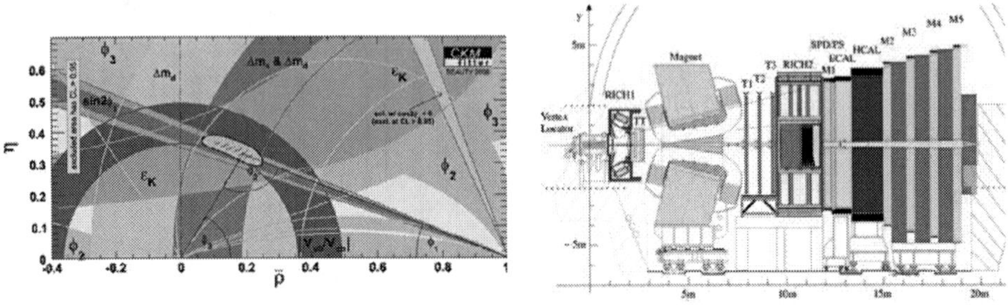

FIGURE 5: a) The impact of B physics on the unitarity triangle
b) The LHCb detector dedicated to B physics at LHC.

The LHCb [15] experiment (Figure 5b) will soon complement and extend the studies made at B factories. This unique detector will probe CP violation in the D0 system at the 10^{-3} level and flavour physics in the quark sector will continue to be an important research direction. Among the key elements of the detector are the Si vertex detector that will measure charged particle tracks, and the RICH detectors, that will identify particles. The shopping list of LHCb includes:

- *Bd and Bs mixing effects (including CPV) in tree decays*
- *CPV in pure hadronic $b \to s$ penguin transitions*
- *Direct CPV (angle γ) in $B \to DK$ tree decays*
- *CPV in $B \to h+h-$ decays (trees + penguins)*
- *CPV (angle α) in $B \to \rho\pi$, $\rho\rho$ decays (trees + penguins)*
- *$b \to s\gamma$ and $b \to sl+l-$ transitions (electromagnetic penguins)*
- *Pure leptonic B decays*
- *Charm and charmonium*

These measurements will significantly improve the accuracy of the unitarity triangle. With high luminosities LHCb may well succeed in finding decays with rates above the SM predictions thus pointing to phenomena requiring new physics.

CONCLUSIONS

The non-invariance of the fundamental interactions under the discrete symmetries C, P and T has played a central role in the development of the SM. Quantitatively, the measurements of CP violation in several channels, both with kaons and b mesons, is perfectly consistent with the SM, and in particular with one single complex phase δ as the dominant source of CP violation in the quark sector.

The measurement of very rare FCNC decays such as $K^0_L \to \pi^0 \nu \bar{\nu}$, $K^+ \to \pi^+ \nu \bar{\nu}$, or $B_{d,s} \to l^+ l^-$, the detection of new CP-violating phases in heavy-flavour decays, will undoubtedly provide precious information on the effect of these new phenomena on low-energy physics, and perhaps give important constraints on yet unobserved heavy particles.

ACKNOWLEDGMENTS

It was a pleasure to participate to this first UAE-CERN workshop held in Al-Ain in order to help this region of the word deciding upon how to organize their participation in the adventure of finding things yet unknown to mankind. I thank the organizers for their wonderful effort and the invaluable help and kindness that they provided to their guests.

REFERENCES

1. M. Gell-Mann and A. Pais, *Phys. Rev.* ***97****, 1387, (1955).*
2. K. Landé *et al.*, *Phys Rev.* ***103****, 1901 (1956).*
3. D. Neagu *et al.*, *Phys Rev. Lett.* ***6****, 552 (1961).*
4. L. B. Leipuner *et al.*, *Phys Rev.* ***132****, 2285 (1963).*
5. J. H. Christenson, J. W. Cronin, V.L. Fitch and R. Turlay, *Phys Rev Lett. 13,138 (1964).*

6 L. Wolfenstein, *Phys. Rev. Lett.* 13, *562, (1963).*
7 G. D. Barr *et al.*, *Phys. Lett. B* 317, *233, (1993).*
8 L. K. Gibbons *et al.*, *Phys.Rev. Lett.* 70, *1203, (1993).*
9 A. Alavi-Harati *et al.*, *Phys. Rev.* D67, *012005 (2003).*
10 J. R. Batley *et al.*, *Phys. Lett. B* 544, *97, (2002).*
11 Selected theory papers: *http://www.phy.bnl.gov/e949/papers/theory.html.*
12 S. Balev *for NA48 Collaboration., Rencontres de la Vallée d'Aoste, La Thuile. (2006),*
13 J. R. Batley *et al.*, *Phys. Lett. B* 638, *22, (2006).*
14 Proposal to measure the rare decay $K^+ \rightarrow \pi^+ \nu \bar{\nu}$ at the CERN SPS, CERN-SPSC-2005-013, SPSC326.
15 LHCb Home page: http://lhcb.web.cern.ch/lhcb

ALICE: The Quest for 'Primordial' Matter

C. W. Fabjan

CERN, Geneva, Switzerland

Abstract. ALICE — An LHC Ion Collider Experiment — is being prepared to study, in an optimized and dedicated approach, the physics of nuclear matter under extreme conditions of temperature and density. The LHC Ion programme is the logical sequel in the quest to study the novel form of matter, the Quark-Gluon Plasma, the form of matter believed to have existed in our Universe during the first microsecond after the Big Bang. The status of ALICE, its research programme and discovery potential are presented.

Keywords: Quark-Gluon Plasma; Relativistic Ion Collisions; ALICE.
PACS: 12.38Mh; 21.65Qr; 25.75Nq

INTRODUCTION

ALICE — An LHC Ion Collider Experiment — is the LHC experiment dedicated to the study of relativistic heavy ion collisions. It is planned that the LHC will be operated for about one month each year in the 'Ion Collider' mode, in the first instance colliding fully stripped lead nuclei, and later other systems such as pPb or lighter ions. This contribution highlights some features of the ALICE Detector, its measurement capabilities, and provides a glimpse of its very rich physics programme.

HEAVY-ION COLLISIONS AT THE LHC: A NEW REGIME

The study of relativistic heavy-ion collisions has been a very active area of research during the past 30 years or so. Work started at the Bevalac, and then moved later to the AGS at BNL and the SPS at CERN. Most recently, research at the dedicated 'Relativistic Heavy-Ion Collider', RHIC, at BNL has been particularly fruitful. While work at the CERN SPS has revealed a 'New State of Matter', the Au–Au collisions at RHIC at an energy of $\sqrt{s}$ = 200 GeV/N have provided evidence for a strongly interacting plasma with partonic degrees of freedom and with very low viscosity.

CERN will bring its new machine, the Large Hadron Collider, LHC, into operation during 2008. Proton–proton collisions at $\sqrt{s}$ = 14 TeV and nucleus–nucleus collisions at $\sqrt{s}$ = 5.5 TeV/N will open a new era of particle physics. In Pb–Pb collisions an energy density of 10 to 20 GeV/ fm^3 is expected to be reached, two to three times higher than that at RHIC and approximately ten times higher than the critical energy density above which hadronic matter will transform into the Quark–Gluon Pasma

CP1006, *High Energy Physics and Applications*, edited by J. Ellis, S. Nasri, and E. Malkawi

(QGP). The rates of some of the most interesting probes of the QGP will also dramatically increase at LHC energies, see Fig. 1. Charm production will rise by a factor of ~10, Beauty by a factor of ~100 and jet production will increase by several orders of magnitude.

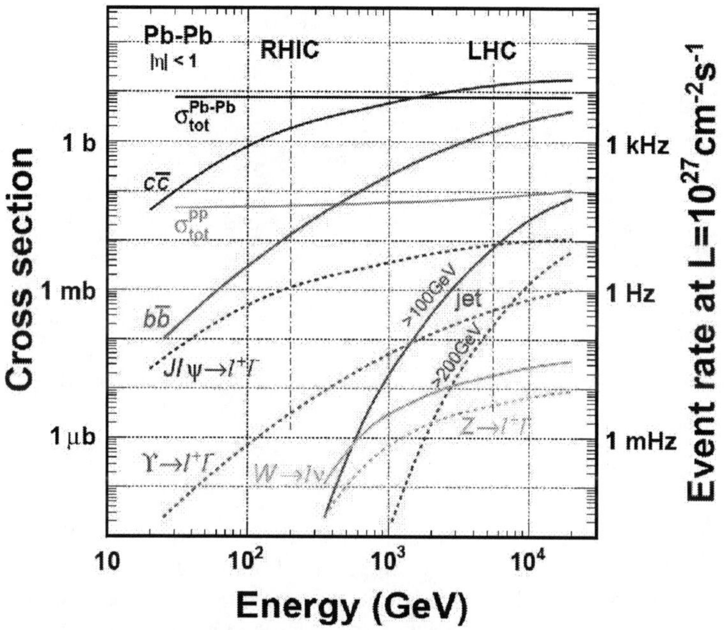

FIGURE 1 : Cross sections and rates from RHIC to LHC.

THE ALICE EXPERIMENT

The ALICE experiment (Ref. [1]) is shown in Fig. 2. It is organized around two magnets. A 0.5 T warm solenoid is the heritage of the LEP era and housed the L3 experiment. In ALICE it accommodates a suite of 'central' detectors, most of which cover 2π in azimuth and approximately two units of rapidity, $|\Delta\eta| \approx 1$.

A particle at mid-rapidity will traverse radially the following detectors:

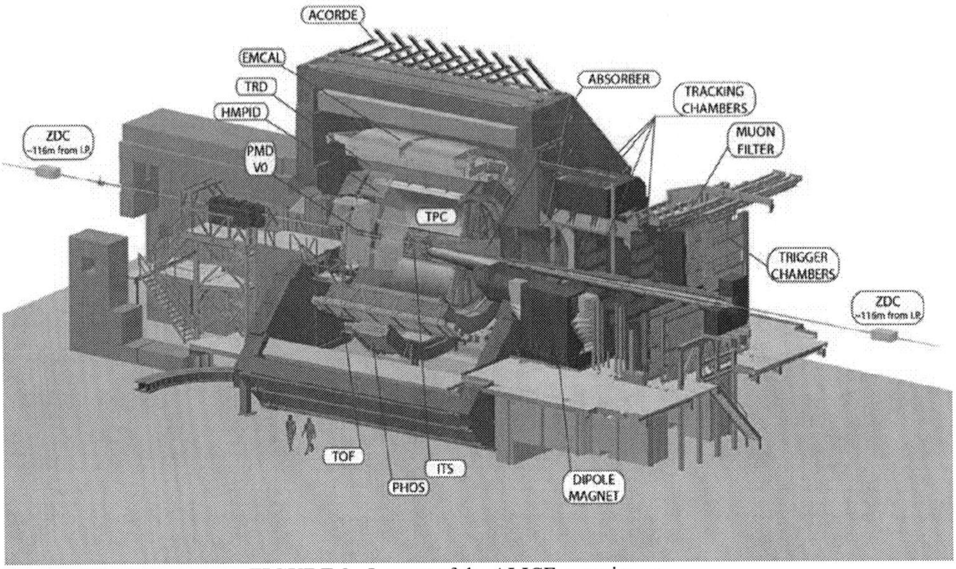

FIGURE 2 : Layout of the ALICE experiment.

i) Inner Tracking System, ITS: This consists of two layers of a Si pixel detector, SPD, with a total of $\sim 9 \times 10^6$ pixels. Following are two layers of a novel Si drift detector, SDD, providing unambiguous space points with 30 μm × 50 μm resolution. Two layers of double-sided Si strip detectors complete the ITS. A noteworthy point is the very light construction. A particle at 90° will see only 7% of a radiation length, implying little multiple scattering and very good secondary vertex resolution (σ (vertex) $\leq$ 50 μm for particles with $p > 1.5$ GeV/c).

ii) Time Projection Chamber, TPC: This is the world's largest TPC with a detector volume of 88 m^3 and a diameter of 5.6 m. The design ensures excellent tracking capability for the highest expected particle densities, $dN/dy \leq 8000$. It is achieved by implementing a very light construction (3% of X_0 for 90° particles), use of a novel detector gas [$Ne/CO_2/N_2$ (81/14.3/ 4.7)] and powerful on-detector electronics. A specially developed ASIC (ALTRO, ALICE TPC Readout chip) allows digital filtering, baseline restoration, and data compression.

iii) Transition Radiation Detector, TRD: A six-layer TRD provides powerful electron/ hadron discrimination and track information. Most importantly, it also provides within 6 μs a trigger decision on high-p electrons and hadrons.

iv) Time-Of-Flight, TOF: Particle identification is an important aid in the study of signals from the QGP. A novel TOF system with $\sigma \approx 50$ ps provides π/K/p identification for $p \leq 4$ GeV/c. This excellent time resolution is achieved by using a novel version of the Resistive Plate Chambers, RPCs. A 250 μm gap is used such that signal fluctuations are minimized; 10 such gaps per detector ensure that particles can be detected with ~ 100 % efficiency.

v) Dedicated central detectors: The central detectors are completed by

a. a 11 m^2 Ring Imaging Cherenkov (HMPID) providing π/K/p identification up to 6 GeV/*c*,

b. a 20 000 $PbWO_4$ crystal calorimeter for high-precision photon and electron identification (PHOS),

c. a large electromagnetic calorimeter (EMCAL) to trigger and measure high-p_t jets.

A large-aperture Muon Spectrometer, covering about 4.0 < y < - 2.8, is arranged around a 0.5T warm dipole magnet. Five tracking stations, using cathode-pad chambers with $\sigma \approx 60$ μm space resolution, provide good momentum resolution. As an example, the Y family will be measured with 1% mass-resolutions. Two planes of RPCs with a dual-threshold discrimination for improved timing resolution ($\sigma < 2$ ns) provide a fast trigger for muons with $p > 1$ GeV/*c*.

A third set of detectors in the 'forward' direction provides specific functionalities:

– Two highly-granular discs of scintillating tiles (V0) provide a fast-charged particle multiplicity detector and beam-gas rejection

– An array of quartz Cherenkov detectors with $\sigma \leq 30$ ps time resolution provide the start signal for the TOF detector

– Forward Multiplicity Detectors (FMDs) provide charged-particle information in the forward rapidity region.

– A Photon Multiplicity Detector (PMD) provides information on photons in the forward direction.

– Zero Degree Calorimeters (ZDCs) at 116 m from the collision point provide an independent trigger on centrality.

The performance of the central tracking system is very powerful. From 100 MeV/*c* up to 100 GeV/*c* the momentum resolution is below 5%. It is also noteworthy that ALICE has efficient particle tracking to below 100 MeV/*c*.

The central detector system also has excellent particle identification:

– π/K/p identification to 6 GeV/*c* with HMPID and TOF; powerful d*E*/d*x* identification with σ(d*E*/d*x*) = 5.5 to 6.5%, depending on particle multiplicity, provides π/K/p information up to 50–60 GeV/*c*;

– electron/ pion discrimination from 1 to > 100 GeV/*c* with the TRD;

– photon identification from 2 to > 100 GeV with PHOS;

– identification based on the reconstruction of secondary vertices (K,Λ,Ξ,Ω,c,b).

PHYSICS HIGHLIGHTS

A detailed discussion of the ALICE physics reach is given in Ref. [2] and [3]. ALICE will contribute to both the pp - programme and the ion programme. The pp programme will provide the necessary reference data to interpret heavy-ion collisions. It will, however, also contribute to specific studies on event properties, particle composition and multiplicity, taking advantage of its unique PID capabilities and its momentum coverage to very low values.

The RHIC research has shown that jets can powerfully probe the properties of the matter produced. This is most clearly revealed in modifications of the jet fragmentation function. The quality of ALICE for such measurements is shown in Fig. 3. The (simulated) measurements (open circles) closely follow the results which would be obtained with a perfect jet-detector (solid line) demonstrating the sensitivity of the measurements.

One of the most striking signatures of the QGP should be the 'melting' of heavy mesons, such as J/ψ and Υ. Despite twenty years of studies, the experimental evidence still does not allow clear conclusions. We expect that the new measurements at the LHC (Fig. 4) together with the RHIC results will provide the solution to this generation-old enigma.

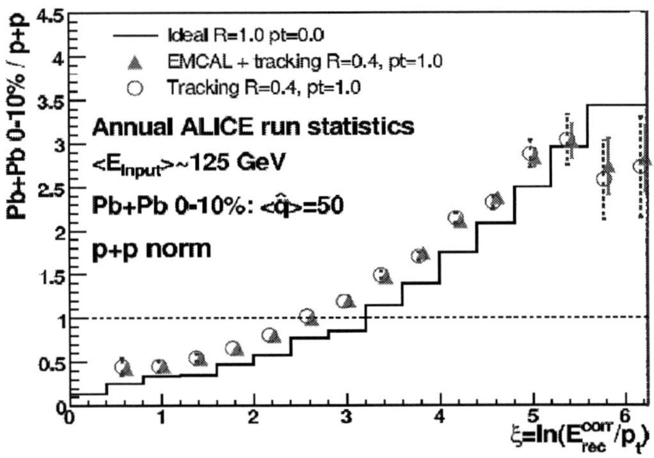

FIGURE 3: Simulated measurement of the jet fragmentation function. Shown is the performance of ideal detectors (black line) as compared to the ALICE performance without and with an electromagnetic calorimeter.

OUTLOOK

ALICE is almost completely installed and currently undergoing commissioning of the complete hardware and software chain, all the way including GRID computing, preparing itself for the exciting years ahead. While the ALICE Collaboration is by now a mature collaboration with more than 1000 members, new groups are still joining, making important contribution to the ALICE programme. Beyond the initial studies we are considering a range of 'upgrades' to adapt ALICE to this rapidly evolving field.

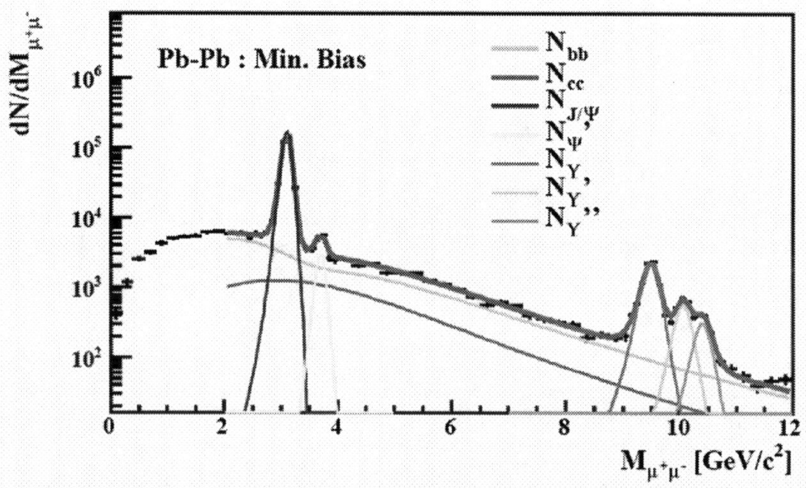

FIGURE 4: Simulated di-muon spectrum obtained in one month of Pb–Pb collisions at nominal luminosity. Together with the RHIC results one expects to understand the 'melting' of the quarkonia states in the QGP, as predicted by QCD.

ACKNOWLEDGMENTS

Discussions with F. Antinori, K. Safarik and J. Schukraft are gratefully acknowledged. The Organizers should be congratulated for having arranged a stimulating meeting, providing the opportunity to discuss research in particle physics and to exchange fruitful ideas on future, increased collaboration. It is hoped that this seminar may be the start of a long-lasting tradition strengthening the collaboration between the Gulf States and CERN.

REFERENCES

1. *ALICE Technical Proposa, CERN/LHCC 95-711,* Geneva: CERN, 1995.
2. F. Carminati et al., *ALICE Physics Performance Report Vol. I., J. Phys. G. Nucl. Part. Phys 30 (2004) 1517.*
3. B. Alessandro et al., *ALICE Physics Performance Report Vol. II, J. Phys. G. Nucl. Phys. 32 (2006) 1295*

The Silicon Inner Tracker of CMS: Construction, Organization and Future Upgrade

Manfred Krammer[1] *on behalf of the CMS Tracker Collaboration*

Institute of High Energy Physics, Austrian Academy of Sciences, Vienna, Austria

Abstract. The silicon Inner Tracker is the innermost detector part of the CMS experiment at CERN. This detector contains about 200 m^2 of precision silicon sensors, fast and radiation hard electronics, an optical readout and many more highly sophisticated components. The total number of sensor elements amounts to 9.3 million strips and 66 millions pixels. The strip detector is completely assembled and has been commissioned at the operating temperature of about $-15°$ C. A collaboration consisting of about 500 members coming from 51 institutes worldwide has designed and constructed this detector in a period of more than 12 years. The different components for the tracker were developed in cooperation with industrial partners. The Inner Tracker was installed in the CMS experiment at the end of 2007. A discussion on the upgrade of CMS and the Inner Tracker has already started in view of an increase in performance of the LHC collider. To cope with the even more demanding requirements expected for the next generation apparatus an extensive R&D period is about to start with many opportunities opening up for new collaboration members to join the effort.

Keywords: LHC, CMS, Silicon detectors
PACS: 29.40.Wk; 29.40.Gx

LAYOUT AND CONSTRUCTION

At CERN, the European Laboratory for Particle Physics in Switzerland, the large hadron collider LHC is under construction. The LHC will produce proton-proton collisions at a centre of mass energy of 14 TeV, the startup of the LHC is foreseen for 2008. To exploit this machine the CMS experiment, a multipurpose experiment, is being build at CERN to take data at the startup of the LHC.

The CMS detector consists of several shells of different detector elements. Particles created in the high energy collisions in the very center of the detector will first traverse the 'Inner Tracker' a system of silicon sensors designed to provide a precise and efficient measurement of the trajectories of charged particles.

This Inner Tracker [1] is geometrically divided into several substructures (see figure 1) the pixel detector very close to the interaction point and the strip tracker consisting of the inner barrel detector (TIB), the inner discs (TID), the outer barrel (TOB) and the two end cap detector systems (TEC). The overall length of the Inner Tracker is 5.4 m with an outer diameter of 2.4 m. The temperature inside the tracker will be adjusted such that the maximum temperature of the silicon sensors will not exceed $-10°$C. Details on the layout and the construction of the pixel and of the strip detector are explained in the

---

1 krammer@hephy.oeaw.ac.at

CP1006, *High Energy Physics and Applications*, edited by J. Ellis, S. Nasri, and E. Malkawi

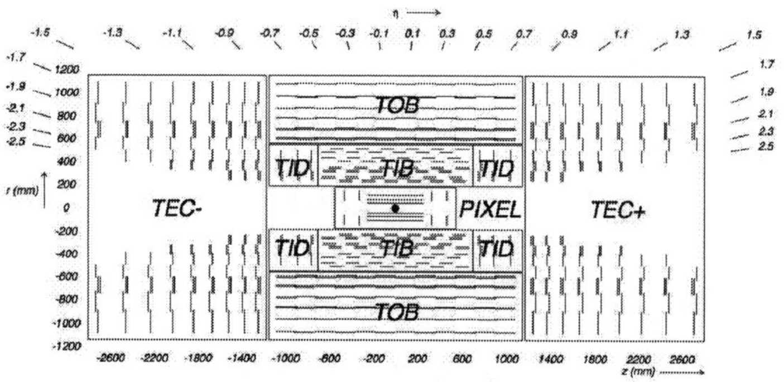

FIGURE 1. Schematic cross section through the CMS tracker. Each line represents a detector module. Double lines indicate back-to-back modules which deliver stereo hits.

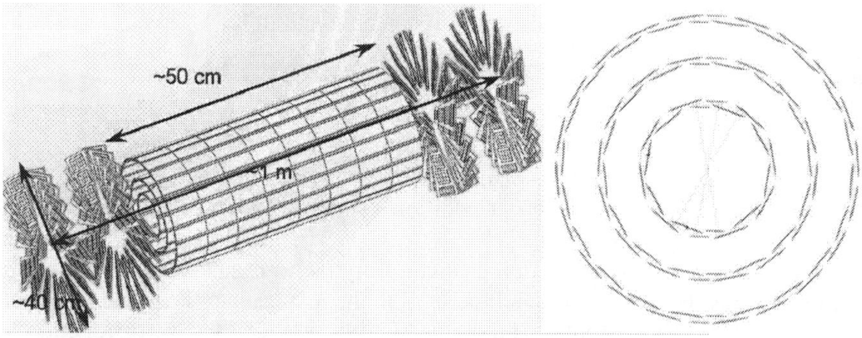

FIGURE 2. View of the pixel detector (left) and cut through the central part (right).

following subsections.

The Pixel Detector

Three cylindrical layers of hybrid pixel detector modules surround the interaction point at radii of 4.4, 7.3 and 10.2 cm. Two discs of pixel modules on each side complement the pixel detector. Figure 2 shows a graphical view of the pixel detector and a cut through the central part.

The pixel detector modules are built as hybrid pixel assemblies containing the components described in the following. The active silicon sensors are realized on high resistance n-substrate with an implanted pn-junction and a pixel cell size of 100x150 μm^2. Indium bumps are deposited onto the sensors for subsequent connection to the readout electronics. The readout electronics consists of custom ASICs fabricated in a commercial 0.25 μm process. Each chip processes the signals from 4160 pixels. Up to 16 chips are bump bonded onto one sensor wafer. On top of the sensor chip assembly is a low

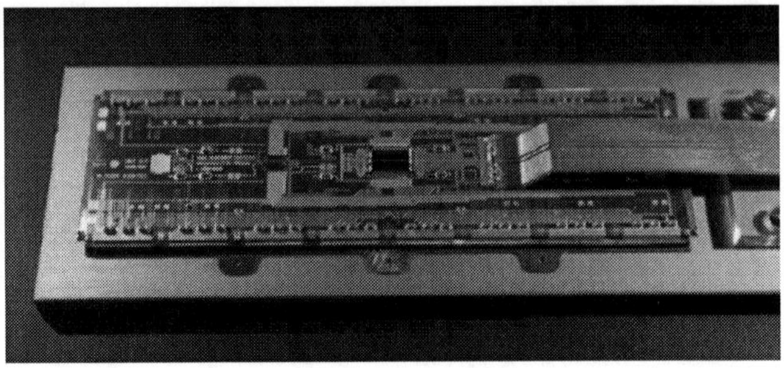

FIGURE 3. A barrel pixel module (66.6 x 26.0 mm^2).

mass multilayer printed circuit board holding an additional control chip and other components. Figure 3 shows a picture of an assembled barrel pixel module. Further details on the technology used for the pixel detector can be found in [2].

The Strip Detector

The strip detector surrounds the pixel detector and adds additional 10 detector layers in the central region (4 TIB, 6 TOB). In addition 3 small and 9 large detectors discs (TID and TEC) are located on either side [3].

The basic construction element of the silicon strip tracker is a module. The supporting frame of a module is made of carbon fibre or graphite. Glued onto the frame is a Kapton layer to isolate the frame from the silicon [4] and to provide the electrical connection to the silicon backplane. A ceramic multilayer circuit (hybrid) holds the readout chips and the auxiliary chips. A glass pitch adapter is mounted between the hybrid and the first silicon sensor to match the different pitches of the chips' input pads and of the sensor strips. Wire bond connections between the individual channels of the readout chips and the pitch adapter, between the pitch adapter and the first sensor and between the two sensors provide the electrical connections. The modules of the TIB, the TID and the four inner rings of TEC consist of only one silicon sensor, whereas the modules of the TOB and the three outer rings of TEC hold two sensors. All barrel modules are of rectangular shape. The modules of the discs have a wedge shape in order to form rings. Figure 4 shows a final module of the second ring of the TEC. The first two layers in TIB and TOB, the first two rings in TID and the rings 1, 2, and 5 in TEC are instrumented with double-sided modules. These are made of two independent single-sided modules, mounted back to back and rotated by 100 mrad with respect to each other. Table 1 lists some numbers illustrating the overall dimensions of the Inner Tracker strip detector.

Figure 5 shows a completed half shell of the tracker inner barrel (TIB). The sensitive silicon modules on the inside and outside (not seen in the figure) of the shell overlap such that no gap exists where particles could escape undetected.

One of the two completed tracker end caps (TEC) is shown in figure 6 before being inserted in the tracker support tube.

TABLE 1. Some key parameters of the silicon strip tracker construction.

Area of active silicon	$\approx 200\ \mathrm{m}^2$
Number of silicon sensors	24,244
Different sensor designs	15
Number of modules	15,148
Mechanically different module designs	27
Number of strips	$\approx$ 9,300,000
Number of electronics channels	$\approx$ 9,300,000
Number of readout chips	$\approx$ 73,000
Number of wire bonds	$\approx$ 25,000,000

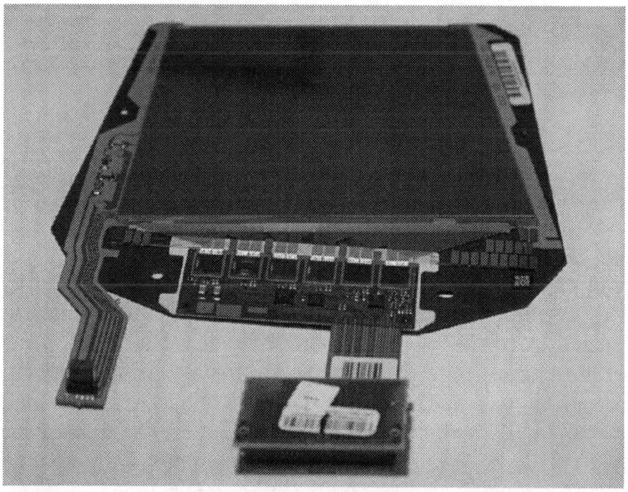

FIGURE 4. A module for the TEC subdetector of the CMS silicon tracker.

FIGURE 5. One half shell of the tracker inner Barrel.

FIGURE 6. A tracker end cap ready for installation.

Commissioning and Installation

The strip tracker was completed at CERN using the tracker integration facility - a clean room with facilities to assemble, connect and operate part of the tracker in turn. The tracker volume was sealed in April 2007 and about 15% of the strip tracker was cabled and operated first at room temperature, and later on at the operating temperature of about -15° C. Trigger scintillators above and below the tracker allowed the selection of cosmic muons traversing the tracker volume. In a period of six months, among many other tests, about 5 million muon tracks were recorded. The full readout chain - including the software for track reconstruction and alignment - was commissioned, tested and optimized. The measurements during this period have shown that 99.8% of all strip tracker channels comply with specifications, and only 0.2% of the channels are defective. Everywhere the efficiency in the tracker is higher than 99.5%. The signal to noise ratio is about 27 for TIB modules and 31 for TOB and TEC modules [5].

The sealed CMS Inner Tracker was finally transported from the integration facility to the experimental area and lifted down into the cavern. On December 15, 2007 the tracker was inserted into the final place inside the experiment CMS. Figure 7 shows a picture taken during the insertion. In the meantime the pixel detector is being completed. After the installation of the beam pipe in CMS, the pixel detector will be inserted into the tracker in April 2008. This will be the final detector part to complete the CMS experiment.

THE CMS TRACKER COLLABORATION

The tracker collaboration is formed by 51 institutes originating from 9 different countries: Austria, Belgium, Finnland, France, Germany, Italy, Switzerland, United Kingdom, and the USA. Figure 8 shows the geographical distribution of the CMS tracker

FIGURE 7. Installation of the world's largest silicon tracking detector in the CMS experiment at CERN.

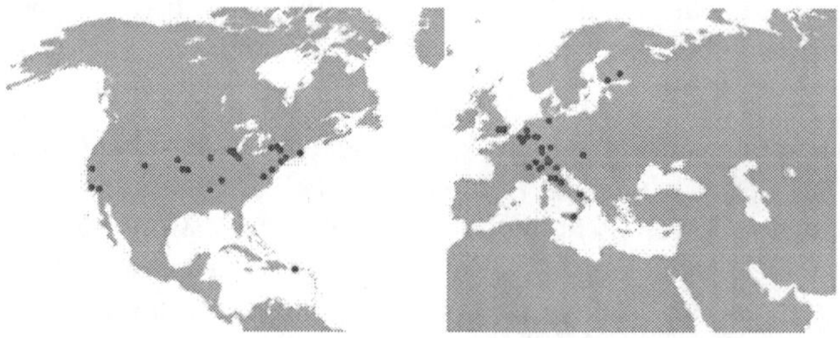

FIGURE 8. Map showing the location of tracker institutes in the US and Europe.

institutes in Europe and in the US. The collaboration consists of about 500 physicists and academic engineers. During design, construction and commissioning of the CMS tracker each institute has taken the responsibility for well defined tasks. Most of the work on hardware and software was carried out at the home institutes by staff members and students of various levels of expertise. Although such a distributed production scheme did require significant additional resources for coordination and logistics the benefit of in-house work is appreciated by all institutes.

FUTURE UPGRADE

The present CMS tracker is designed to be operated for a lifetime of 10 years until about 2018 at the LHC design luminosity of 10^{34} $cm^{-2}s^{-1}$. The lifetime of the silicon detector is limited by the radiation induced change of the doping concentration in the substrate and by the subsequent need for an increase of the operating voltage. The planning of the CERN management foresees to upgrade the LHC machine in steps towards a possible luminosity of about 10^{35} $cm^{-2}s^{-1}$ at around 2018. At this luminosity the particle flux emerging from the interaction point exceeds the capabilities of the present CMS tracker and of the present CMS trigger system. As a consequence of both the predicted end of the tracker lifetime and of the need for an improved tracker to operate at the increased luminosity, CMS is planning to construct a new tracker for 2018 [6].

The challenges for the construction of this new tracker are manifold:

- The upgraded tracker has to withstand enormous radiation levels up to about $1.3x10^{16}$ n/cm^2 for the innermost detector layers. In this respect the development of new sensor materials is required.
- The particle fluxes are expected to be an order of magnitude larger compared to the present tracker. In order to keep the channel occupancy low, smaller cell sizes are needed what leads to an increase in channel numbers.
- The new Inner Tracker has to provide data for the level 1 trigger decision logic. This is mandatory for CMS to reduce the event rate to be processed.
- Despite the higher number of channels the power dissipation must not be increased compared to the present tracker.
- The amount of material in the tracker volume, e.g. in the support and cooling structures, has to be significantly reduced.

The research and development effort for this new tracker has just started with the formation of several working groups. In the next five years intense research on the critical items should result in the optimal approach for the construction. The R&D phase will then be followed by about two years of qualification and further three years of construction. The main targets to be studied are the sensor technologies, ASIC development, simulation, mechanical issues, power distribution, triggering, cooling, and related subjects. Now, while the research proposals for all these fields are being discussed and defined, it is an ideal time for new groups to join the CMS tracker upgrade project.

REFERENCES

1. CMS Tracker TDR, CERN/LHCC 98-6 CMS TDR 5 (1998) and CMS TDR Addendum, LHCC 2000-016 (2000).
2. The CMS pixel detector, A. Dominguez, Nucl. Instr. and Meth. A 581 (2007) 343.
3. Construction of the CMS silicon strip tracker, G. H. Dirkes, Nucl. Instr. and Meth. A 581 (2007) 299.
4. The Silicon Sensors for the Compact Muon Solenoid Tracker - Design and Qualification Procedure, J.-L. Agram et al., Nucl. Instr. and Meth. A 517 (2004) 77.
5. CMS Tracker commissioning and first operation experience, Ch. Delaere, CMS CR-2008/002, Proceedings of the 16th International Workshop on Vertex Detectors, Lake Placid, September 2007.
6. CMS Expression of Interest in the SLHC, CERN/LHCC 2007-014 , LHCC-G-131, March 15, 2007.

Extra Dimensions and New Vector Bosons Searches in CMS

W. Emam

LLR, CNRS/IN2P3, Ecole Polytechnique, Palaiseau, France

Abstract.
This paper addresses a series of searches for signals from extra dimensions and new vector bosons in the CMS experiment. A brief review of the CMS discovery potential of these signatures in different luminosity scenarios is presented.

Keywords: LHC, CMS, Extra Dimensions, Extra Gauge Bosons

INTRODUCTION

Many extensions of the Standard Model are motivated by the hierarchy problem between the electroweak scale (100 GeV) and the Plank scale ($10^{(}19)$ GeV) where the relativistic and quantum effects are important for gravity. Several solutions have been proposed. The most promising ones are the Supersymmetry and Extra dimensions models. The models of extra dimensions (ED) have very rich low-energy phenomenology that originates from the KK spectrum of various particles propagating in ED.

The CERN Large Hadron Collider (LHC) is currently being installed and commissioned in the 27 km ring previously used for the LEP e+e- collider. This machine will push up the high energy frontier by almost one order of magnitude, providing pp collisions at $\sqrt{s} = 14$ TeV with expected luminosity of 10^{34} cm^{-2}s^{-1}. The LHC collaborations have studied the main models introducing ED. In this paper we discuss various experimental signal and the discovery potential of the CMS experiment to observe these signatures considered in the ADD, RS and TeV^{-1} models.

MODELS OF EXTRA DIMENSIONS

ADD Model

The existence of several large spatial extra-dimensions in which gravity propagates was first proposed by Arkani- Hamed, Dimopoulos and Dvali (ADD) [1]. In this ED scenario, the observed three-dimensional space is a 3D-brane embedded in a higher dimensional space, the extra spatial dimensions (the bulk) being orthogonal to the 3D-brane. The SM fields are localized on the 3D-brane and only the gravitons may propagate in extra dimensions. The phenomenological consequence of this scenario is the appearance of, massive Kaluza-Klien (KK) modes of graviton exist and lead to violations from SM predictions in the TeV scale region. These gauge boson modes are

CP1006, *High Energy Physics and Applications*, edited by J. Ellis, S. Nasri, and E. Malkawi

spin 2 particles, couple to the SM particles, and have masses given by

$$M_n^2 = M_0^2 + n^2/R^2 \tag{1}$$

where M_0 is the mass of the zeroth mode, corresponding to the SM fields, n is the mode number and R is the compactification radius.

These KK-gravitons can be directly produced at the LHC in $q\bar{q} \to gG$, $gq \to qG$, $gg \to gG$ processes or observed through virtual KK-graviton exchange. In these cases experimental signals for the ADD scenario might be found in di-jet, di-lepton, and di-photon mass spectra as well as missing energy distributions.

TeV^{-1} size extra dimentions

In addition to the large extra dim, smaller ones are introduced (of TeV^{-1} size) in which gauge bosons can propagate while fermions are confined on the 4-dim branes [2]. The phenomenological consequence of this scenario is the appearance of the KK tower of states of massive gauge boson fields. Therefore, the main discovery channel comes from the observation of the decay of a heavy particle into an electron and muon pair (Z_{KK} decays) or lepton an missing E_T carried out by neutrinos (W_K decays).

Randall-Sundrum Model RS

In the model by Randall and Sundrum [3], gravity is located on a (3+1)-dimensional brane, the Planck brane, that is separated from the SM brane in a fifth dimension with warped metric. In the simplest version of this model gravitons are the only particles that can propagate in the extra dimension. The gravitons appear as towers of Kaluza-Klein excitations with masses and widths determined by the two free parameters of the model, namely, the mass of the first excited mode of the graviton, M_1, and the dimensionless coupling to the SM fields, $c = k/M_{Pl}$.

Two theoretical constraints exist on these two parameters. The first is a limit on the curvature $k < M$, where $M \sim M_{Pl}$ is the fundamental five-dimensional Planck scale. The second constraint on the gravity scale, $\Lambda < 10$ TeV, assures that no new hierarchy appears between M_{EW} and Λ. On the other hand, the experimental constraint on RS graviton production comes from the Tevatron Run II, which excludes a graviton mass up to 535 GeV=c^2 for c = 0.1 [4].

SEARCH FOR ED AT CMS

The LHC will provide sufficiently high energy in the center of mass that allows the CMS experiment to search for the direct production of new heavy resonances. Over the past decade, several studies were already performed on the search for these heavy resonances in various experimental channels [5]. The results presented here focus on dilepton and diphoton final states.

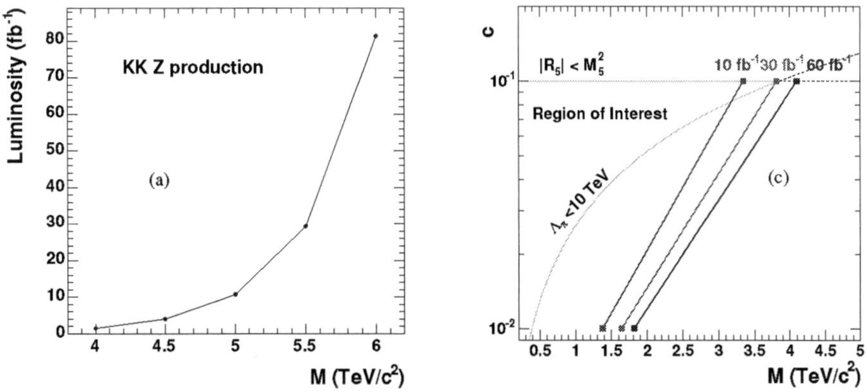

FIGURE 1. Electron pair production is considered. *Left* 5σ discovery limit of the resonance mass for KK excitation of Z production. *Right* 5σ discovery plane for RS graviton production as function of the coupling parameter c and the graviton mass.

The e^+e^- decay channel provides a clean signature in the CMS detector. The presence of a heavy particle would be detected in CMS by the observation of a resonance peak in the dielectron mass spectrum over the Drell-Yan process $pp \rightarrow \gamma/Z \rightarrow e^+e^-$ which constitutes the main Standard Model background. Figure 1 gives the 5σ discovery limit of the resonance mass for KK excitation of Z production (*left*) and the 5σ discovery plane for RS graviton production as function of the coupling parameter c and the graviton mass (*Right*).

Two classes of large extra dimensions models, RS and ADD, are studied with $\mu^+\mu^-$ final states. The results for different values of integrated luminosity are summarised in Fig. 2. The figure shows that the CMS experiment can observe a RS graviton with mass up to 2.3 TeV/c^2 with an integrated luminosity of $\int Ldt = 1$ fb^{-}1 if the coupling c is equal to 0.1.

Due to the spin-2 nature of the graviton its branching ratio to di-photons is roughly twice that of a single di-lepton channel. In addition, the study of the RS graviton decaying into the two photons is particularly interesting as the detection of such few TeV/c^2 mass resonance in such channel together with its observation in the di-lepton channel will sign a RS graviton, distinguishing it from a Z' production. The discovery region in the plane of the coupling parameter c and the graviton mass is shown in Fig. 3.

EXTRA GAUGE BOSONS

A possible way of resolving the inherent problems of the standard model is by extending the gauge sector of the theory. New heavy gauge bosons are predicted in many extensions of the standard model [6]. For instance, in grand unified theories heavy partners of the electroweak bosons generally appear; the left-right symmetric model is a SO(10) GUT

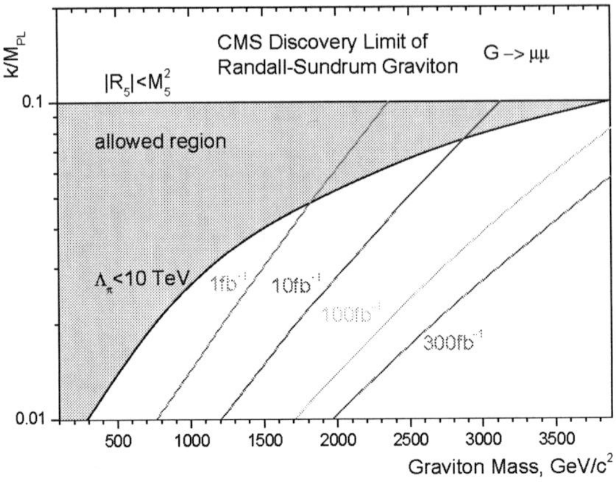

FIGURE 2. Muon pair production is considered. 5σ discovery plane for RS graviton production as function of the coupling parameter c and the graviton mass M_G.

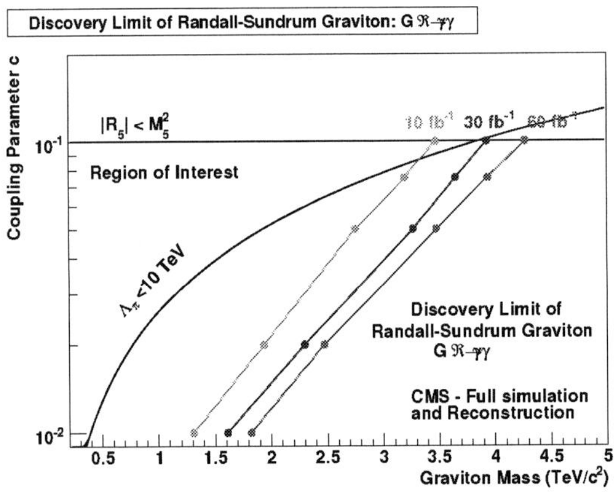

FIGURE 3. Photon pair production is considered. 5σ discovery limit of the resonance mass for RS graviton production as function of the coupling parameter c and the graviton mass.

extension of the SM, postulating the existence of a right-handed version of the weak interaction as well as an additional Z boson. In addition, the sequential standard model (SSM), where the couplings to quarks and leptons are as in the SM, serves as a good benchmark for comparisons of results.

The search for Z' bosons in the di-electron and di-muon channels can be considered one of the early physics channels to be looked at at CMS. Even with small amounts of data, a di-lepton resonance can clearly be separated from the SM background (mostly Drell-Yan production of lepton pairs) for masses above the present experimental limits from the Tevatron, which are of the order of 1 TeV. The expected discovery reach in the di-electron channel as a function of the Z' mass and the integrated luminosity

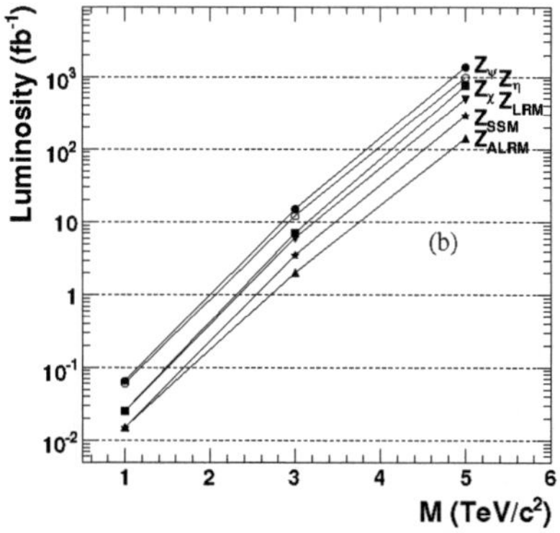

FIGURE 4. 5σ discovery limit in the integrated luminosity and Z' mass plane.

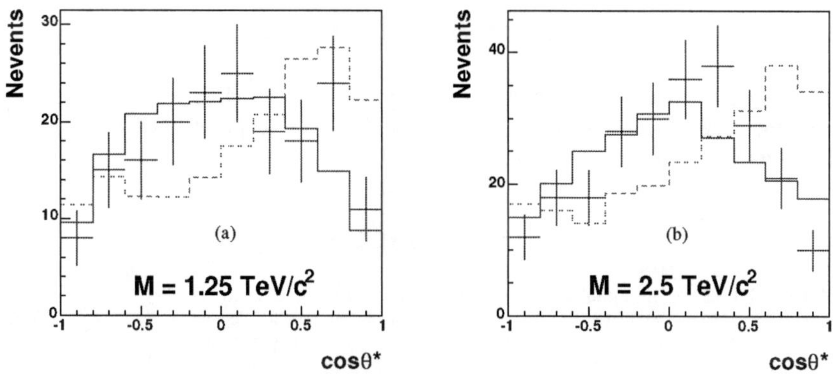

FIGURE 5. Distributions of $\cos\theta$ for RS graviton production (full curves) for M = 1.25 and 1.5 TeV/c^2 for $c = 0.01$ and for Drell-Yan production (dotted curves).

for different Z' models is shown in Fig. 4. The results in the di-muon channel are comparable.

On the other hand, RS gravitons with spin 2 can be distinguished from the Standard Model background and Z' bosons with spin 1 using the distribution of the $\cos\theta$ variable, computed as the cosine of the polar angle between the electron and the boost direction of the heavy particle in the latter rest frame. The $\cos\theta$ distributions for graviton production with M = 1.25 and 2.5 TeV=c^2 are presented in Fig. 5 and Fig. 6 for an integrated luminosity of 100 fb^{-1} [7]. The measurement of the forward-backward asymmetries of leptonic decay products, both at the resonance peak and off the peak, is also known to be a powerful tool to identify Z' [8].

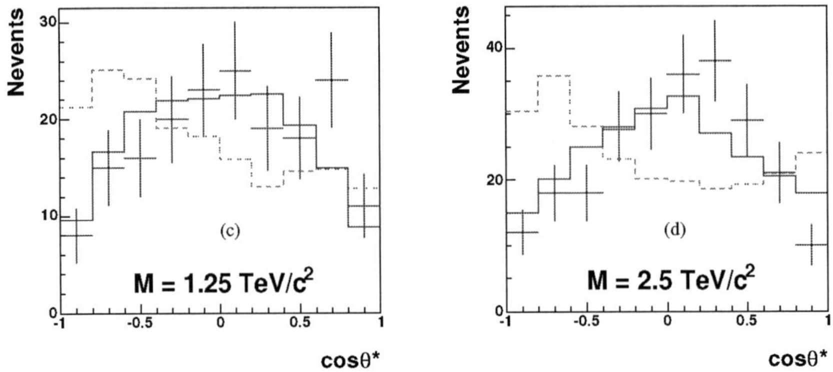

FIGURE 6. Distributions of $\cos\theta$ for RS graviton production (full curves) for M = 1.25 and 1.5 TeV/c^2 for $c = 0.01$ and for Z' production (dotted curves).

ACKNOWLEDGMENTS

It is a pleasure to thank C. Charlot, L. Dobrzynski, and P. Mine for very useful discussions and assistance. We acknowledge the financial support provided by Ecole Polyetechnique and the CNRS, France.

REFERENCES

1. N. Arkani-Hamed, S. Dimopoulos and G. R. Dvali, Phys. Lett. B **429**, 263 (1998).
2. I. Antoniadis, Phys. Lett. B **377**, 246 (1990), I. Antoniadis and K. Benakli, Phys. Lett. B **326**, 69 (1994), I. Antoniadis, K. Benakli, and M. Quiros, Phys. Lett. B **331**, 313 (1994),
3. L. Randall and R. Sundrum, Phys. Rev. Lett. **83**, 3370 (1999).
4. CDF Collaboration, Phys. Rev. Lett. **87**, 131802 (2001).
5. C. Collard and M.-C. Lemaire, **CMS Note 2004/024**. M.-C. Lemaire, V. Litvin, H. Newman, **CMS Analysis Note 2005/032**. M.-C. Lemaire, J.-P. Pansart, **CMS Note 2006/051**. B. Clerbaux, C. Collard, T. Mahmoud, P. Mine, **CMS Note 2006/083**. R. Cousins, J. Mumford, V. Valuev, **CMS Note 2006/062**.
6. J. L. Hewett and T. G. Rizzo, Phys. Rep. **183**, 139 (1989), M. Cvetic and S. Godfrey, arXiv:hep-ph/9504216, A. Leike, Phys. Rept. **317**, 143 (1999), J. Erler and P. Langacker, Phys. Lett. B **456**, 68 (1999), T. G. Rizzo, arXiv:hep-ph/0610104, W. M. Yao *et al.* [Particle Data Group], J. Phys. G **33**, 1 (2006), E. Boos, A. Djouadi, M. Muhlleitner and A. Vologdin, Phys. Rev. D **66**, 055004 (2002), W. Emam and S. Khalil, Eur. Phys. J. C, 10.1140/epjc/s10052-007-0411-7, arXiv:0704.1395 [hep-ph], M. Carena, A. Daleo, B. A. Dobrescu and T. M. P. Tait, Phys. Rev. D **70**, 093009 (2004).
7. B. Clerbaux, C. Collard, T. Mahmoud, P. Mine, **CMS Note 2006/083**.
8. R. Cousins, J. Mumford, V. Valuev, **CMS Note 2005/022**.

Particle Detectors: Research and Development at CERN

C. W. Fabjan

CERN, Switzerland

Abstract. Over the past 15 years a worldwide Detector R&D Programme has made the LHC experiments possible. These experiments operate at a new level of event rate and detection capabilities. Based on these advances, Detector R&D is continuing at CERN in close collaboration with University and Research Institutes. Several main directions are being pursued for solid-state and gaseous tracking devices, advanced crystal and noble liquid calorimetry, particle identification methods, and advanced signal-processing techniques. This effort is directed towards experiments at even higher collision rates at the LHC, the requirements for the next generation of linear electron–positron colliders and for applications outside particle physics, such as medical diagnostics instrumentation. We shall illustrate this challenging, stimulating and creative programme with examples and show how these developments are taking place in close collaboration between CERN and institutions around the globe.

Keywords: Particle Detector; Particle collider; Detector Application
PACS: 29.40.Cs; 29.40.Gx; 29.40.Mc; 29.40.Wk

INTRODUCTION

A major R&D programme during the 90s has provided the basis for the LHC experiments. While these experiments are currently undergoing intensive commissioning in time for first beams later in 2008, a detector R&D programme is being started to cover certain future needs, in particular for:

– the luminosity upgrade of the LHC to 10^{35} cm^{-2} s^{-1},
– the requirement at a future e^+e^- linear collider,
– certain applications outside particle physics, e.g. medical imaging.

The CERN Council has recently recognized the need for this R&D initiative by allocating significant additional resources. As in the past, such R&D is being carried out in international collaborations. The non-CERN-based collaborators usually contribute with work both at their home institutions and at CERN, e.g., participating in the beam tests.

I shall comment on a few initiatives which may be of particular interest to institutions of the Gulf countries.

CP1006, *High Energy Physics and Applications*, edited by J. Ellis, S. Nasri, and E. Malkawi

R&D IN MICROPATTERN GAS DETECTORS (MPGD)

This relatively recent category of gaseous detectors holds great promise for both high-rate and high-precision applications. The concept is shown in Fig. 1. The thin wire of a traditional proportional chamber is replaced by a matrix of microstructures at which a sufficiently high electric field causes electron multiplication. Two structures are most frequently used: A wire mesh in the 'Micromegas' technology or an insulating foil, covered with a metal film, in which holes have been chemically machined, Gaseous Electron Multiplier (GEM). These detectors provide 'pixilated' readout with ≤ 50 μm space resolution and rate capabilities up to 10^6 particles/mm^{2}s. Apart from the amplification layer ('GEM foil'), such detectors are relatively easy to construct with tooling available in most physics departments.

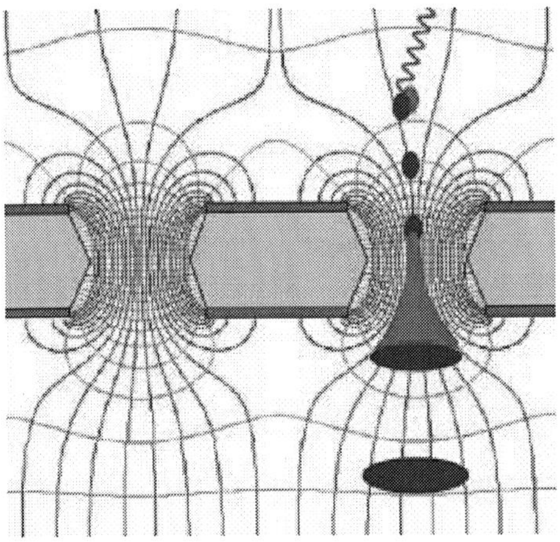

Figure 1: Cross section through the amplification foil ('GEM foil') of a GEM detector. Across the orifice an electric field, strong enough to cause electron multiplication, amplifies the ionization charge. The amplified charge induces a signal in suitably cathode planes with small pick-up electrodes.

These technologies are being considered for the upgrade of the ATLAS Muon Spectrometer and for the readout of Time Projection Chambers for linear colliders. The importance of this detector is underlined by the recent formation of a worldwide R&D collaboration (RD51) aiming to study

- detector physics: performance optimization; gas studies; ageing; discharge prevention; radiation hardness.
- system aspects: new technologies for large surface detectors; new detector geometries; resistive electrodes; new materials;
- detector simulation;
- electronics development;

References [1] and [2] provide further information.

The collaboration with CERN provides access to valuable technologies and infrastructures:

- it has the capability to produce critical components (GEM foil);
- it has the infrastructure for component quality control, clean-rooms, bonding facilities;
- it provides test beams and irradiation facilities.

Personally, I would like to add that participation in such a programme has many attractive features:

- these detectors will be part of future experiments; it is 'cutting edge' R&D;
- the University–CERN collaboration takes advantage of the respective strengths and is well adapted to University infrastructures;
- needed simulations can be carried out at Universities;
- MPGDs have a wide range of applications, from particle and nuclear physics to material testing and medical imaging;
- this R&D programme is an excellent and stimulating training ground for students.

R&D IN RESISTIVE PLATE CHAMBERS (RPC)

While not part of the present CERN R&D programme it is worth while to mention this because:

- it is being considered for linear colliders detectors to 'time-stamp' events with 50 ps resolution;
- RPCs are the basis for a number of innovative projects in Europe and USA to study extensive air showers, initiated by very high energy cosmic rays;
- these projects involve in some cases High School students in an outstanding didactic approach to bring physics to the class rooms, addressing topical physics issues;
- involvement of High Schools is realistic, because simple RPCs are relatively easy to construct; electronics is relatively inexpensive.

Reference [3] provides in-depth information.

R&D ON RADIATION-HARD SEMICONDUCTOR DETECTORS

This work is motivated by the expected LHC luminosity upgrade to 10^{35} cm^{-2} s^{-1} which will lead to fluences in the inner trackers ranging from 10^{14} to 10^{17} neutron-equivalent particles/cm^2.

Research is proceeding along several lines [4]:

- novel silicon materials: The process of silicon crystal growth and the oxygen content have a significant impact on the radiation hardness;
- device engineering of detectors: Construction and doping are being optimized for increased radiation resistance;

- novel detector geometries: One example, Fig. 2, shows a particularly promising geometry of electrodes, giving both adequate detector thickness for full detection efficiency and short drift distances which reduce the effect of radiation damage.

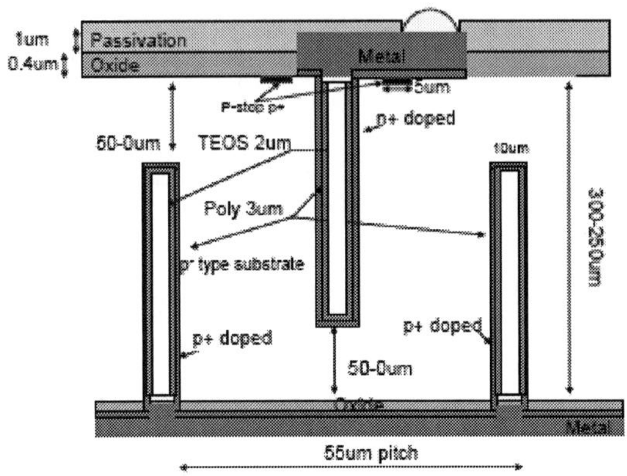

Figure 2. Cross section through a 3-dimensional silicon detector showing the transverse charge collection and resulting in reduced collection paths.

One very important spin-off of this research is the development of a silicon pixel detector with an associated readout chip, the 'MEDIPIX', aimed primarily at medical imaging applications [5].

Perhaps the most striking recent development is the marriage of an MPGD with a pixel readout chip; in this concept the electronic avalanche generated in the MPGD is directly sensed by the input transistor of the readout chip, opening new possibilities of 3-dimensional imaging [6].

R&D FOR CRYSTALS

Particle physicists have a long tradition of developing crystals for precision electron and photon spectroscopy. The most recent success story is the R&D on $PbWO_4$, carried out for the LHC experiments ALICE and CMS.

Work is continuing in two directions:

- calorimetry at a future linear collider will require better-then-ever jet detection qualities, requiring equal response to hadrons, electrons, and neutrons. One line of the R&D studies the use of fibres — scintillating, 'Cherenkov and neutron-sensitive — with selective readout. Typical of these studies is again work organized in a large, international collaboration, including also scientists and engineers outside particle physics. Industry is a needed, very important, collaboration partner;
- new crystals, optimized for medical imaging applications, e.g., for positron emission tomography (PET). The talents combined for high energy physics applications are equally optimal for medical imaging work. As one example, the

'Crystal Clear' Collaboration has been very successful in both domains [7]. Interestingly enough, the readout techniques of crystals for high energy physics and the next generation of PET instrumentation are surprisingly similar: technology transfer at its best!

PERSONAL CONCLUDING REMARK

Detector R&D is an excellent way to enter into experimental nuclear and particle physics. Collaborations between University Institutes and CERN have a long and fruitful tradition, where both partners usually complement each other very well. Last but not least, it is a stimulating, effective, and instructive way to reach out to and involve students.

ACKNOWLEDGMENTS

The organizers should be congratulated for a stimulating, fruitful meeting. I hope that this is the beginning of a long-lasting tradition.

REFERENCES

1. http://mpgd.web.cern.ch/mpgd
2. http://indico.cern.ch/conferenceDisplay.py?confId=473
3. http://www.tifr.res.in/~rpc2007/
4. http://rd50.web.cern.ch/RD50/
5. http://medipix.web.cern.ch/MEDIPIX/
6. http://www.nikhef.nl/~d90/gossip/
7. http://crystalclear.web.cern.ch/crystalclear/CCCdataref.htm

PARALLEL SESSION CONTRIBUTIONS

The 2HDM Inspired by Low Energy QCD Linear Sigma Model with Two Nonets

Sherif Moussa[1a], Salah Nasri[b], Renata Jora[c] and Joesph Schechter[c]

[a]*Department of Mathematical Science, College of Science, UAEU, Al-Ain 17551, UAE*
[a]*Department of Mathematics, Faculty of Science, Ain Shams University, Cairo, 11566 Egypt*
[b]*Department of Physics, College of Science, UAEU, Al-Ain 17551, UAE*
[c]*Department of Physics, Syracuse University, Syracuse, NY 13244-3011, USA*

Abstract. We show that the *K*-matrix unitarization scheme used to calculate the $\pi\pi$ scattering amplitude in the framework of $\sigma\pi$ linear sigma model can be employed to obtain information about WW scattering amplitude in the Standard Model where the physical scalar particles are too heavy. We show how the above procedure can be followed in comparing the Higgs sector of two Higgs doublet models with the linear sigma model with two nonets in the limit of very heavy physical scalar particles.

Keywords: Linear Sigma Model, *K* Matrix Unitarization, $\pi\pi$ scattering, 2HDM
PACS: 12.60.Fr, 14.70 Fm, 14.80 Bn, 14.80 Cp

INTRODUCTION

The two Higgs doublet models (2HDM) [1] are of the simplest and straightforward extensions of the Standard Model (SM). These models have rich phenomenology since they allow for spontaneous CP violation [2], and provide FCNC [3]. Moreover they provide an interesting phenomenology because of their contents of charged, scalars and pseudo scalars Higgs. The most general 2HDM has 14 parameters. This number can be reduced by imposing particular symmetries like U(1) or Z_2 symmetry.

If experiments at LHC didn't detect SM Higgs, then Higgs could be very heavy. In that case it is useful to look for an approach which can treat non-perturbative Higgs sector. The *K* matrix unitarization scheme [4] which describes the non-perturbative scalar sector of QCD can be implemented to describe the non-perturbative Higgs sector. The comparison between the Higgs sector of the SM and the scalar sector of the linear sigma model has been treated in [5].

We are interested in looking for an approach which can treat non-perturbative Higgs sector in models with two Higgs doublets. We show that the approach followed in [4] could be relevant here.

[1] Speaker. On leave from Department of Mathematics, Faculty of Science, Ain Shams University, Cairo, Egypt.

CP1006, *High Energy Physics and Applications*, edited by J. Ellis, S. Nasri, and E. Malkawi

CASE OF ONE HIGGS DOUBLET

The low energy QCD Lagrangian in terms of the pion and sigma is

$$L=\frac{1}{2}(\partial_\mu \vec{\pi}.\partial^\mu \vec{\pi}+\partial_\mu \sigma.\partial^\mu \sigma)-V(\sigma,\vec{\pi}) \ , \tag{1}$$

where,

$$V(\sigma,\vec{\pi})=-\frac{\mu^2}{2}(\sigma^2+\pi^2)+\frac{\lambda}{2}(\sigma^2+\pi^2)^2 . \tag{2}$$

Here the sign of λ is chosen to ensure the spontaneous symmetry breaking. We write the Higgs doublet of the SM as

$$\Phi=\begin{pmatrix} i\pi^+ \\ \dfrac{\sigma-i\pi^0}{\sqrt{2}} \end{pmatrix} , \tag{3}$$

Then the Lagrangian in Eq. (1) takes the form

$$L=\partial_\mu \Phi^+.\partial^\mu \Phi+\mu\Phi^+\Phi-\lambda(\Phi^+\Phi)^2 \ , \tag{4}$$

which is the electroweak Higgs sector. Here we introduce the notation $v=\langle\sigma\rangle$. We should note that Eq. (4) can be considered as a scaled version of Eq. (1), since the vacuum expectation value (vev) of sigma is $\langle\sigma\rangle=0.0926\ GeV$, while the electroweak vev is $\langle\Phi\rangle=0.246\ TeV$.

K-matrix Unitarization Scheme and Linear Sigma Model

We consider the SU(2) linear sigma model in Eq. (1) to calculate the scattering amplitude of two pions in the s-wave isosinglet channel. The contributions for $\pi^+\pi^- \rightarrow \pi^+\pi^-$ are coming from the contact term, s-channel and the crossed Higgs boson exchange. The $I=J=0$, partial wave amplitude at tree level can be calculated as

$$\left[T_0^0\right]_{tree}(s)=\frac{m_{\sigma b}^2}{32\pi F_\pi^2}[-10+\frac{4m_{\sigma b}^2}{s}\ln(\frac{m_{\sigma b}^2+s}{m_{\sigma b}^2})]+\frac{3m_{\sigma b}^4}{16\pi F_\pi^2(m_{\sigma b}^2-s)} . \tag{5}$$

Here $F_\pi=\sqrt{2}\langle\sigma\rangle$ is the pion decay constant and $m_{\sigma b}$ is tree level sigma bare mass. The normalized amplitude T_0^0 is related to the partial wave S-matrix by $S_0^0=1+2iT_0^0$. But the last formula works well at threshold, and can't be trustable at even several hundred MeV above the threshold. Moreover the amplitude in Eq.(5)

diverges at $s = m_{\sigma b}^2$. Hence we will employ the K-matrix regularization which let's use the unitary form

$$S_0^0(s) = \frac{1+iT_0^0}{1-iT_0^0}. \tag{6}$$

Physical Mass of Sigma

Now the question one might ask is: how non-perturbative is the linear sigma model when it is employed as an approximation to low energy QCD? The answer depends on the value of the coupling constant λ in Eq. (2). From the values of $m_{\sigma b}$, obtained from fitting with experimental data, and $\langle\sigma\rangle$ we could see that $\lambda >> 1$. Thus the theory lies outside the perturbative region. In non-perturbative regime one could expect the physical sigma mass is different from its *bare* mass. To investigate this we look for the sigma pole position in the partial wave amplitude $T_0^0(s)$. The pole position z_0 is then given as the solution of the equation

$$(m_{\sigma b}^2 - z_0)(1 - i\frac{m_{\sigma b}^2}{32\pi F_\pi^2}[-10 + \frac{m_{\sigma b}^2}{z_0}\ln(\frac{m_{\sigma b}^2 + z_0}{m_{\sigma b}^2})]) - i\frac{3m_{\sigma b}^4}{16\pi F_\pi^2} = 0. \tag{7}$$

For treating both the low energy QCD as well as the standard electroweak situation it is convenient to introduce the scaled quantities

$$\overline{m} = \frac{m_{\sigma b}}{F_\pi} = \frac{m_{\sigma b}}{\sqrt{2}v}; \quad \overline{z_0} = \frac{z_0}{F_\pi^2} = \frac{z_0}{2v^2}. \tag{8}$$

Writing Eq.(7) in terms of $\overline{m}$ and $\overline{z_0}$, then solving the resulting equation numerically we get $\overline{z_0} \approx \frac{352\pi^2}{3\overline{m}^2} - 8\pi i$. The quantity $\sqrt{\mathrm{Re}(z_0)}$ can be interpreted as a sigma physical mass. Cleary $\sqrt{\mathrm{Re}(z_0)}$ depends on $m_{\sigma b}$. There is a maximal limit for the physical mass around $m_{\sigma b} = 0.74\ GeV$ which correspond to the value $\overline{m} = 5.65$. Increasing $\overline{m}$ beyond this point will decrease the physical mass.

From Linear Sigma Model to SU(2) Electroweak Model

The discussion of $\pi\pi$ scattering amplitude above can be used to treat the high energy scattering of the longitudinal components of W and Z bosons in the SM electroweak theory by using the Goldstone boson equivalence theorem [6]. It states that " at high energy the amplitude of longitudinally massive gauge bosons equal to the amplitude of the Goldstone boson that was eaten by the gauge boson. Here $W_L^+W_L^-$ and $\pi^+\pi^-$. So

$$amp(W_L^+W_L^- \to W_L^+W_L^-) = amp(\pi^+\pi^- \to \pi^+\pi^-)$$
$$amp(W_L^+W_L^- \to Z_L Z_L) = amp(\pi^+\pi^- \to \pi^0\pi^0) \tag{9}$$

In calculating the amplitudes on the right hand side, we should use the value $v = \langle \sigma \rangle = 0.246\, TeV$. To compare the unitarized longitudinal electroweak vector boson scattering with the analogue $\pi\pi$ scattering we plot the relation between "physical" Higgs mass $\sqrt{\mathrm{Re}(Z_{pole})}$ and the bare Higgs mass.

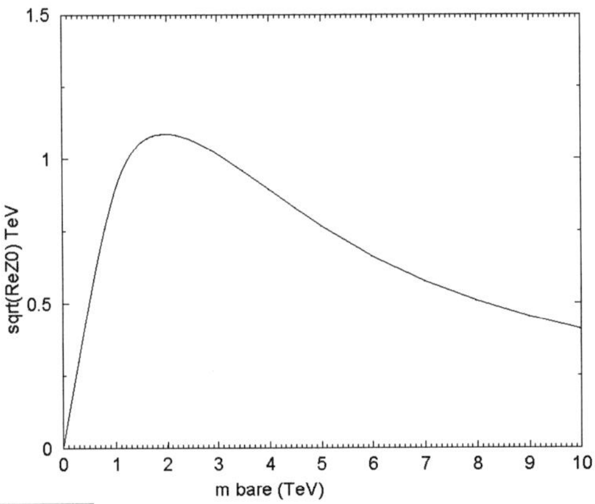

FIGURE 1. $\sqrt{\mathrm{Re}(Z_{pole})}$, of the unitarized amplitude, as a function of bare Higgs mass.

We recall that the "QCD" sigma has a bare mass $m_{\sigma b} \approx 850\, MeV$ which gives $\overline{m} = 6.49$. This value of $\overline{m}$ corresponds to a bare Higgs boson mass value of $m_{\sigma b} = 2.26\, TeV$. At that value the measure of the physical Higgs boson mass $\sqrt{\mathrm{Re}(z_0)}$ would be about $1.1\, TeV$. The most important point of this analysis is that we can trust the applicability of K matrix unitarization model for the Lagrangian given in Eq.(4) to the bare Higgs masses in at least $2 TeV$ region. In other words the same model with the same scaled parameters agrees with experiment there in a different context. Since the present status of the Higgs sector of the standard electroweak theory may change if future experiments reveal evidence for new physics just beyond the standard model, it seems worthwhile to have some confidence in an approach to a possibly non-perturbative Higgs sector. In particular as the QCD analog two-sigma model shows, the K-matrix unitarization method can be expected to work in the case of more than one Higgs particle even in non-perturbative regions of parameter space. For example, the introduction of the σ' resonance would scale to a 2 Higgs model describing physics up to about 3.3 TeV in some electroweak theory.

CASE OF TWO HIGGS DOUBLETS

The most general form for the potential of 2HDM is

$$\begin{aligned} V(\Phi_1, \Phi_2) = m_{11}|\Phi_1|^2 + m_{22}|\Phi_2|^2 + \lambda_1|\Phi_1|^4 + \lambda_2|\Phi_2|^4 + \lambda_3|\Phi_1|^2|\Phi_2|^2 + \\ [\beta_1\Phi_1^*\Phi_2 + h.c.] + [\beta_2(\Phi_1^*\Phi_2)^2 + h.c.] + [\beta_3(\Phi_1^*\Phi_2)|\Phi_1|^2 + h.c.] + \\ [\beta_4(\Phi_1^*\Phi_2)|\Phi_2|^2 + h.c.] + [\lambda_4(\Phi_1^*\Phi_2)(\Phi_2^*\Phi_1)] \end{aligned} \quad (10)$$

It has 14 parameters, 6 real and 4 complex, with two types of dimensions. Some symmetry can be imposed to reduce the number of parameters. For example if we imposed Z_2 symmetry $\Phi_1 \to \Phi_1, \Phi_2 \to -\Phi_2$, terms contains odd power of Φ_2 will vanish. Imposing Z_2 will avoid FCNC. To see the similarity between the Higgs sector of 2HDM and the scalar sector of low energy QCD linear sigma model with two nontes we assume that

$$\Phi_i = \begin{pmatrix} i\pi_i^+ \\ \dfrac{\sigma - i\pi_i^0}{\sqrt{2}} \end{pmatrix}, \qquad M_i = \sigma_i I + i\vec{\tau}\cdot\vec{\pi}_i, \quad i = 1, 2. \quad (11)$$

Cleary $M_i = (i\tau_2\Phi_i^*\ \Phi_i)$. Moreover M_i transforms as $M_i \to U_L M_i U_R^+$. The QCD Lagrangian that describes σ_i and $\vec{\pi}_i$ is

$$L = \frac{1}{2}Tr(\partial_\mu M_1.\partial^\mu M_1^+) + \frac{1}{2}Tr(\partial_\mu M_2.\partial^\mu M_2^+) - V(M_1, M_2), \quad (12)$$

where

$$\begin{aligned} V(M_1, M_2) = \alpha_1 I_1 + \alpha_2 I_2 + \alpha_3 I_3 + \alpha_4 I_1^2 + \alpha_5 I_2^2 + \alpha_6 I_3^2 + \\ + \alpha_7 I_1 I_2 + \alpha_8 I_1 I_3 + \alpha_9 I_2 I_3 \end{aligned}, \quad (13)$$

and

$$\begin{aligned} I_1 = \frac{1}{2}Tr(M_1 M_1^+) = \sigma_1^2 + \vec{\pi}_1^2,\ I_2 = \frac{1}{2}Tr(M_2 M_2^+) = \sigma_2^2 + \vec{\pi}_2^2, \\ I_3 = \frac{1}{2}Tr(M_1.M_2^+) = \sigma_1\sigma_2 + \vec{\pi}_1\vec{\pi}_2 \end{aligned}. \quad (14)$$

Imposing Z_2 symmetry $M_1 \to M_1, M_2 \to -M_2$ to the potential in Eq. (13) will reduce the number of parameters into 6 parameters. It is worthy to note that Z_2 symmetry corresponds to Parity conservation in QCD. In that case the potential in Eq.(13) takes the form

$$V(M_1, M_2) = \alpha_1 I_1 + \alpha_2 I_2 + \alpha_4 I_1^2 + \alpha_5 I_2^2 + \alpha_6 I_3^2 + \alpha_7 I_1 I_2, \quad (15)$$

The two nonets here could be $(\sigma, \vec{\pi}) = (\sigma, \pi_1, \pi_2, \pi_3)$ and $(\eta, \vec{a}) = (\eta, a_1, a_2, a_3)$. In that case we have $\Phi_2 = \begin{pmatrix} -ia^+ \\ \dfrac{\eta + ia^0}{\sqrt{2}} \end{pmatrix}$, and $I_3 = -\vec{\pi}\cdot\vec{a} + \sigma\eta$ to ensure parity conservation.

Applying the minimization conditions we can get the following:

When $\langle\eta\rangle = 0$: we have

$$\langle\pi_i\rangle = 0;\ \langle a_i\rangle = 0; \langle\sigma^2\rangle = -\frac{\alpha_1}{\alpha_4}, \tag{16}$$

The masses are given by

$$\begin{gathered} \left\langle \frac{\partial^2 V}{\partial\sigma^2} \right\rangle = 8\,\alpha_4 \langle\sigma\rangle^2;\ \left\langle \frac{\partial^2 V}{\partial\eta^2} \right\rangle = 2\alpha_2 + 2\,(\alpha_6 + \alpha_7)\langle\sigma\rangle^2; \\ \left\langle \frac{\partial^2 V}{\partial\pi_i \partial\pi_j} \right\rangle = 0; \left\langle \frac{\partial^2 V}{\partial a_i \partial a_j} \right\rangle = 2\delta_{ij}\,(\alpha_2 + \alpha_7)\langle\sigma\rangle^2 \end{gathered} \tag{17}$$

The model in Eq.(15), when $\langle\eta\rangle = 0$, contains two neutral CP even scalars σ and η (σ corresponding to the standard model Higgs), a charged pair a^+ and a^-, one neutral CP odd scalar a^0 and the three pseudo Goldstone bosons π^+, π^- and π^0.

It is straight forward to get the unitrized amplitude and the total cross section:

$$amp(W_L^+ W_L^- \to W_L^+ W_L^-) = \frac{x[x - i(1+2x^2)]}{(1+x^2)(1+4x^2)},\ \text{where } x = \frac{s}{16\pi v^2} \tag{18}$$

$$\sigma(W_L^+ W_L^- \to W_L^+ W_L^-) = \frac{1}{v^2}\frac{x[x^2 + (1+2x^2)^2]}{(1+x^2)^2(1+4x^2)^2} \tag{19}$$

The Plot of the total cross section for the case of $\langle\eta\rangle = 0$ is given below.

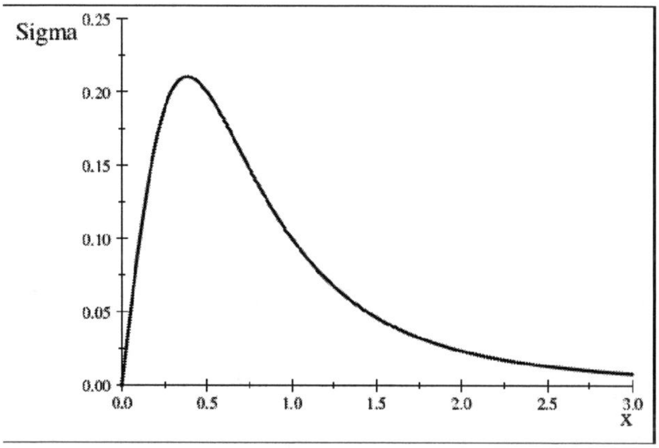

FIGURE 3. The total cross section (in units of nanobar) versus the square energy (normalized by $16\,v^2$)

In the case of $\langle\eta\rangle \neq 0$, more terms will contribute in the scattering amplitudes and this work is still under progress. Moreover we consider also the case with now Z_2 symmetry.

CONCLUSION

We showed that how we can calculate WW scattering amplitude in the SM where the physical scalar particles are very or too heavy. The K matrix unitarization scheme can be also used in treating scattering of heavy scalars in models with 2 Higgs doublets. We could say that Linear Sigma model with two nonets could be an effective lagrangian for the 2HDM where the masses of physical scalars are very or two heavy.

ACKNOWLEDGMENTS

S. Moussa would like to thank Prof. J. Schechter, department of Physics at Syracuse University for his kind hospitality to initiate this work. Also he thanks department of Physics at UAEU for inviting him to give a talk at UAEU-CERN workshop 2007.

REFERENCES

1. D. Toussaint, Phys. Rev. D **18,** 1626 (1978); and H. Georgi, Hadar J. Phys. **1,** 155 (1978).
2. T. D. Lee, Phys. Rev. D **8,** 1226 (1973); G.C. Branco and N. Rebelo, Phys. Rev. D **22,** 2901 (1980).
3. S. I. Glashow and S. Weinberg, Phys. Rev. D **15,** 1958 (1977).
4. N.N. Achasov and G.N. Shestakov, Phys. Rev. D **49,** 5779 (1994).
5. A. Abdel Rehim et.al., Phys. Rev. D **68,** 013008 (2003).
6. J.M. Cornwall, D. N. Levin, and G. Tiktopoulos, Phys. Rev. D **16,** 1519 (1977).

Double Higgs production in the Two Higgs Doublet Model at the linear collider

Abdesslam Arhrib*, Rachid Benbrik† and Cheng-Wei Chiang**

*Department of Physics and Center for Mathematics and Theoretical Physics, National Central University, Chungli, Taiwan 320, R.O.C and Département de Mathématiques, Faculté des Sciences et Techniques B.P. 416 Tanger, Morocco.
†Department of Physics, Chung Yuan Christian University, Chungli, Taiwan 320, R.O.C.
**Department of Physics and Center for Mathematics and Theoretical Physics, National Central University, Chungli, Taiwan 320, R.O.C and Institute of Physics, Academia Sinica, Taipei, Taiwan 115, R.O.C.

Abstract. We study double Higgs-strahlung production at the future Linear Collider in the framework of the Two Higgs Doublet Models through the following channels: $e^+e^- \to \Phi_i\Phi_j Z$, $\Phi_i = h^0, H^0, A^0$. All these processes are sensitive to triple Higgs couplings. Hence observations of them provide information on the triple Higgs couplings that help reconstructing the scalar potential. We discuss also the double Higgs-strahlung $e^+e^- \to h^0h^0Z$ in the decoupling limit where h^0 mimics the SM Higgs boson.

Keywords: Higgs boson, two Higgs doublet model, decoupling limit
PACS: 12.60.-t, 13.66.Fg, 13.85.Lg

1. In order to establish the Higgs mechanism for the electroweak symmetry breaking, we need to measure the Higgs couplings to fermions and to gauge bosons as well as the self-interaction of Higgs bosons. Such measurements, if precise enough, can be helpful in discriminating between models through their sensitivity to quantum corrections, particularly in specific scenarios such as the decoupling limit where Higgs couplings mimic the Standard Model (SM) ones.
In the recent years there has been growing interest in the study of extended Higgs sectors with more than one Higgs doublet [1]. The simplest extension is the Two Higgs Doublet Model (2HDM); and such a structure is required for the Minimal Supersymmetric Standard Model (MSSM). The 2HDM as well as the MSSM predict in their spectrum two CP even h^0, H^0, one CP odd A^0 and a pair of charged Higgs $H^\pm$.

In case of a discovery of the Higgs boson at the Large Hadron Collider (LHC), it may be possible to measure its couplings to gauge bosons and fermions at a certain precision. It is well known that physics at the LHC and the e^+e^- International Linear Collider (ILC) will be complementary to each other in many respects. In many cases, the ILC can significantly improve the LHC measurements.

There have been several studies dedicated to triple Higgs couplings both in the SM and beyond for linear collider (for a review, see Ref [2]). Despite the smallness of its cross section, it has been shown in Ref. [3] that for the process of $e^+e^- \to ZHH$ with $H \to b\bar{b}$ in the SM, the irreducible background from electroweak and QCD processes can be suppressed down to manageable levels by using kinematics cuts.

In the present study, we report on the triple Higgs couplings as well as the double

CP1006, *High Energy Physics and Applications*, edited by J. Ellis, S. Nasri, and E. Malkawi

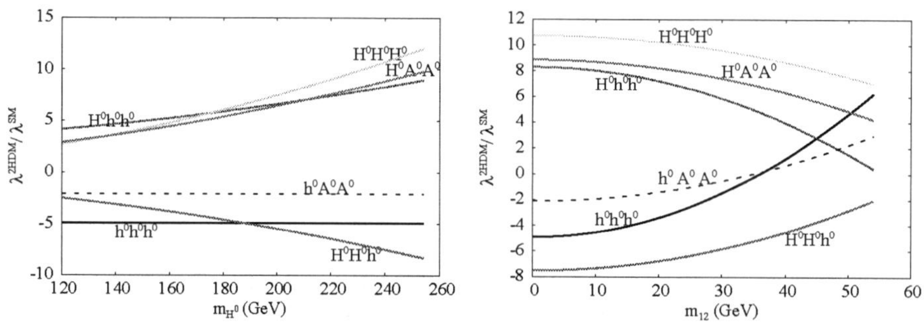

FIGURE 1. The 2HDM tree level self couplings $\lambda^{2HDM}_{h_i h_j h_k}$ normalized to the SM self coupling λ^{SM}_{hhh}: (left) as a function of m_{H^0} with $m_{12} = 40$ GeV, (right) and as a function of m_{12} with $m_{H^0} = 240$GeV. The others parameters are $m_{h^0} = 120$ GeV, $m_{H^\pm} = 140$ GeV, $m_{A^0} = 150$ GeV, $\tan\beta = 10$ and $\sin\alpha = 0.8$

Higgs-strahlung production $e^+e^- \to \Phi_i\Phi_j Z$, $\Phi_i = h^0, H^0, A^0$ in the general 2HDM, as measurements of these processes can shed some light on the 2HDM triple Higgs couplings (see Ref. [4] for a complete review).

2. First, we discuss the behavior of triple Higgs couplings as a function of the 2HDM parameters. The 2HDM parameters are the four physical masses: $m_{h^0,H^0,A^0,H^\pm}$, $\tan\beta$ which is the ratio of the vacuum expectation values of the Higgs doublet, α the mixing angle between the CP even states and m_{12} dimension-two term which violates the discrete symmetry $\Phi_i \to -\Phi_i$ softly.

We will show that varying the scalar parameters within the allowed range can still make the triple Higgs couplings larger than the SM triple Higgs coupling $\lambda^{SM}_{h^0h^0h^0} = -3em^2_{h^0}/(2m_W s_W)$ by several orders of magnitude. These 2HDM triple Higgs couplings can also be large compared to the MSSM triple Higgs couplings. The reason for this is that in the MSSM, supersymmetry imposes restrictions on the quartic couplings λ_i by relating them to the gauge couplings.

In Fig. 1, we illustrate the sizes of the triple Higgs couplings as a function of CP-even Higgs mass m_{H^0} (left plot) and as a function of the discrete symmetry breaking parameter m_{12} (right plot). One can see from these plots that the sizes of triple Higgs couplings normalized to the SM triple Higgs coupling lie between -10 and 12 for all allowed values of m_{H^0} and m_{12}. The variation over m_{H^0} can change the size of these couplings by about one order of magnitude.

The dependence of the triple Higgs couplings on m_{12} is also important (see the right plot of Fig. 1). However, due to the perturbativity and vacuum stability conditions [5], m_{12} is constrained to be less than 55 GeV. We would like to mention also that the 2HDM triple Higgs couplings are enhanced with more than 50 times with respect to SM one in the large $\tan\beta$ limit [4].

3. In this section, we will discuss the production cross section of the following processes: $e^+e^- \to h^0h^0Z$, $e^+e^- \to h^0H^0Z$, $e^+e^- \to H^0H^0Z$ and $e^+e^- \to A^0A^0Z$. Those processes are sensitive to the triple Higgs couplings $h^0h^0h^0$, $h^0H^0H^0$, $H^0h^0h^0$, $h^0A^0A^0$ and $H^0A^0A^0$ which have been studied in the previous section. The complete calculation

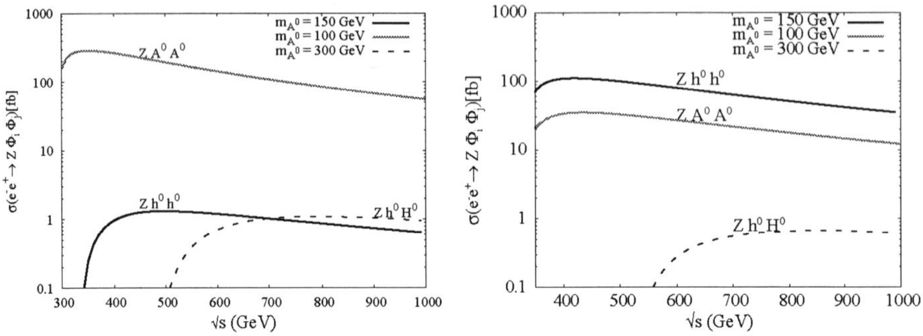

FIGURE 2. Total tree level cross section of $e^+e^- \to Z\Phi_i\Phi_i$ in 2HDM as a function of $\sqrt{s}$ with $m_{h^0} = 120$ GeV, $m_{H^\pm} = 140$ GeV, $m_{12} = 0$ GeV, $\tan\beta = 10$ for different values of m_{A^0}. The left plot uses $m_{H^0} = 200$ GeV and $\sin\alpha = 0.6$. The right plot uses $m_{H^0} = 240$ GeV and $\sin\alpha = 0.8$.

of cross sections is done with the help of the packages FeynArts, FormCalc [6].

Before discussing our numerical results, it is worth pointing out that the following results are valid for both 2HDM-I and 2HDM-II since they involve only the Higgs coupling and the gauge coupling to Higgs or to e^+e^-. In fact, In 2HDM-I, the quarks and leptons couple only to one of the two Higgs doublet exactly the same as in the SM. In 2HDM-II, one of the two Higgs fields couples only to down-type quarks (and charged leptons) and the other one only couples to up-type quarks (and neutral leptons) in order to avoid the problem of large flavor-changing neutral currents (FCNC's). We would like to mention here, that the 2HDM-I can lead to a fermiophobic Higgs boson h^0 in the limit where $\cos\alpha = 0$ [1], and the dominant decay mode for the lightest Higgs boson in this model is $h^0 \to \gamma\gamma$ or $h^0 \to WW$, depending on its mass. As we will see later, this fermiophobic limit can lead to distinctive signature for the double Higgs-strahlung.

In Fig. 2, we show the total cross sections for the neutral modes Zh^0h^0, Zh^0H^0, ZH^0H^0 and ZA^0A^0 as a function of the center-of-mass energy $\sqrt{s}$ for different values of m_{A^0}. As explained in the figure caption, we select specific $\sin\alpha$ values to optimize the total cross sections. For the Zh^0h^0 and ZA^0A^0 modes, the total cross section is maximized for $\sqrt{s} \approx 350-400$ GeV. In the case of Zh^0h^0 with $m_{h^0} = 120$ GeV, the maximum of the total cross section is reached for $m_{A^0} = 150$ GeV, $m_{H^0} = 240$ GeV, and $\sin\alpha = 0.8$ (right panel), and is slightly larger than 100 fb. This enhancement for $e^+e^- \to Zh^0h^0$ is due to the threshold effect ($H^0 \to h^0h^0$ is open). In the ZA^0A^0 mode, the cross section is maximal for $m_{A^0} = 150$ GeV, $m_{H^0} = 200$ GeV, and $\sin\alpha = 0.6$ and is about 300 fb. The cross section of the heavy mode ZH^0H^0 is very small and is not shown; whereas the cross section of the mixed mode ZH^0h^0 is mild, ≈ 1 fb, in both cases.

Fig. 3 shows the behaviors of $e^+e^- \to Zh^0h^0$ and $e^+e^- \to ZA^0A^0$ (left plot) and the branching ratios of the CP-even H^0 (right plot) as a function of m_{H^0} and for m_{h^0}, $m_{H^\pm}$, $m_{A^0}, m_{12} = 120, 250, 150, 0$ GeV, respectively, $\tan\beta = 5$, $\sin\alpha = 0.9$ and $\sqrt{s} = 500$ GeV. To compute the cross sections, we have summed the continuum part $\sigma(e^+e^- \to ZSS)$ and resonant part generated in the chain $e^+e^- \to ZH^0 \to ZSS$ via the resonant H^0 Higgs-strahlung, with $S = h^0$ or A^0. It is obvious that the cross sections of $e^+e^- \to Zh^0h^0$ and $e^+e^- \to ZA^0A^0$ reach their maxima near the threshold regions of $H^0 \to h^0h^0$ and

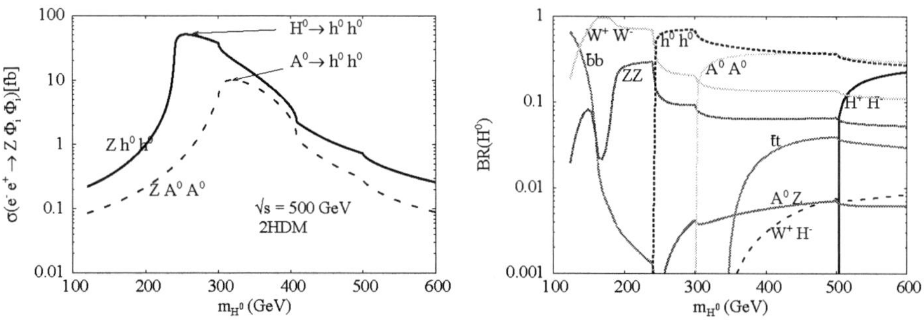

FIGURE 3. Left plot: total cross sections for Zh^0h^0 and ZA^0A^0 continuum and resonant production: $\sigma(e^+e^- \to Zh^0h^0) + \sigma(e^+e^- \to ZH^0) \cdot \mathrm{BR}(H^0 \to h^0h^0, A^0A^0)$ in 2HDM as a function of m_{H^0}. Right plot: branching ratios of H^0. Here $m_{h^0} = 120$ GeV, $m_{H^\pm} = 250$ GeV, $m_{A^0} = 150$ GeV, $m_{12} = 0$ GeV, $\tan\beta = 5$, and $\sin\alpha = 0.9$.

$H^0 \to A^0A^0$.

In the right panel of Fig. 3, we show the decay branching ratios of H^0. The $b\bar{b}$ mode dominates below the WW threshold, while the WW mode takes over for $m_{H^0} \gtrsim 2M_W$. Above the h^0h^0 threshold, the decay $H^0 \to h^0h^0$ is open and dominates over the others at the order of 70%. For $m_{H^0} \gtrsim 300$ GeV, $H^0 \to A^0A^0$ is open and has also a substantial branching ratio of $\mathcal{O}(20\%)$, reducing the branching ratio of $H^0 \to h^0h^0$ to the level of less than 20%.

As shown in the left plot of Fig. 3, there are three kinks occurring when $m_{H^0} = 2m_{h^0}, 2m_{A^0}, 2m_{H^\pm}$ and corresponding to the opening of the $H^0 \to h^0h^0$, $H^0 \to A^0A^0$ and $H^0 \to H^\pm H^\mp$ modes, respectively. The cross sections increase by about 20 fb if h^0h^0 and A^0A^0 are produced in $e^+e^- \to ZH^0 \to Zh^0h^0$ and $e^+e^- \to ZH^0 \to Zh^0h^0$ via resonant Higgs-strahlung of H^0.

In the general 2HDM, the double Higgs-strahlung processes $e^+e^- \to Zh^0h^0$, $e^+e^- \to ZA^0A^0$ and $e^+e^- \to Zh^0H^0$, the dominant final states depend on how h^0, A^0 and H^0 decay. In the case where h^0 and A^0 are not heavy ($\lesssim$ 125 GeV), h^0 and A^0 will decay to $b\bar{b}$ or $\tau^+\tau^-$, and the final states for Zh^0h^0 and ZA^0A^0 would be $Z4b$, $Z2b2\tau$ or $Z4\tau$. In the 2HDM, it may be possible that, if kinematically allowed, A^0 decays into Zh^0. In this case, the final state for ZA^0A^0 would be $ZA^0A^0 \to 3Zh^0h^0 \to 3Z4b$. While in the fermiophobic limit of 2HDM-I, one can obtain a very clear signal of $4\gamma + X$ from the Zh^0h^0 mode. For the Zh^0H^0 mode, the signal depends on how H^0 decays. In 2HDM, the heavy Higgs H^0 can decay to W^+W^- or, if kinematically allowed, to h^0h^0. In the latter case, we have $ZH^0h^0 \to Zh^0h^0h^0 \to Z6\gamma$ with $6\gamma + X$ as a distinctive signature.

4. Let us now discuss $e^+e^- \to h^0h^0Z$ in the decoupling limit. In the 2HDM, the decoupling limit generally refers to the case when all the scalar masses except one formally become infinite and the effective theory is just the SM with one doublet (see [7] for recent discussions). In this case, the CP-even h^0 is the lightest scalar particle while the other Higgs particles H^0, A^0 and $H^\pm$ are extremely heavy. Using purely algebraic arguments at the tree level, one can derive that the main consequences of the decoupling limit are $\cos(\beta - \alpha) \to 0$, and the CP-even h^0 of the 2HDM and the SM Higgs h_{SM} have

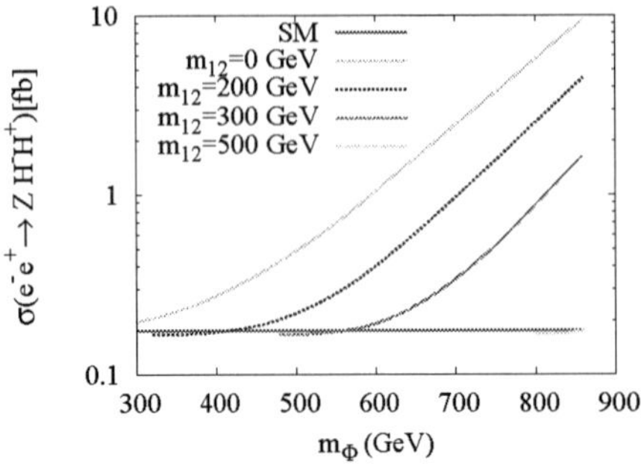

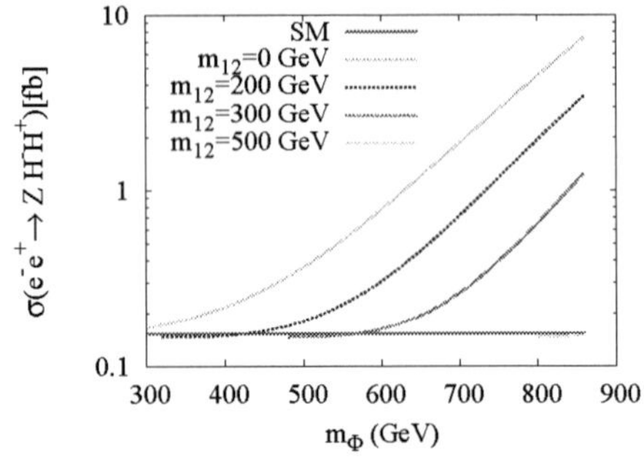

FIGURE 4. $\sigma(e^+e^- \to Zh^0h^0)$ in units of fb as a function of M_Φ at $\sqrt{s} = 500$ GeV (left) and $\sqrt{s} = 800$ GeV (right) for different values of m_{12}. For both plots, $m_{h^0} = 120$ GeV, $\tan\beta = 2$ and $\cos(\beta - \alpha) = 0$.

similar tree-level couplings to gauge bosons and fermions as well [7, 8].

Several studies have been carried out looking for non-decoupling effects in Higgs boson decays and Higgs self-interactions. Large loop effects in $h^0 \to \gamma\gamma$, $h^0 \to \gamma Z$ and $h^0 \to bb$ have been pointed out for the 2HDM [9] and may provide indirect information on the Higgs masses and the involved triple Higgs couplings such as $\lambda_{h^0H^+H^-}$, $\lambda_{h^0H^0H^0}$, $\lambda_{h^0A^0A^0}$ and $\lambda_{h^0h^0h^0}$. The non-decoupling contributions to the triple Higgs self-couplings $\lambda_{h^0h^0h^0}$ have been investigated in the 2HDM in Ref. [8], revealing large non-decoupling effects.

In this section, we will show that the large non-decoupling effects in $\lambda_{h^0h^0h^0}$ modify the double Higgs-strahlung $e^+e^- \to Zh^0h^0$ cross section and make it larger than the SM expectation. We will focus on the scenario where all the Higgs particles of the 2HDM, except for the lightest CP-even Higgs, are heavy and can escape from detection at the first stage of next generation colliders.

For our calculation of the cross section of $e^+e^- \to Zh^0h^0$ in the decoupling limit, we ignore one-loop corrections to the ZZh^0 coupling and replace the $h^0h^0h^0$ coupling by its effective coupling given in Ref [8]. In fact, it has been shown in Ref. [8] that the non-decoupling effect in ZZh^0 is at a few percent level in the case of $\cos(\beta - \alpha) = 0$.

In Fig. 4, we show the cross section of $e^+e^- \to Zh^0h^0$ in the SM and 2HDM in its decoupling regime as a function of $m_\Phi = m_{A^0} = m_{H^0} = m_{H^\pm}$.
For our choice of $m_{h^0} = 120$ GeV, the SM cross section is tiny, ≈ 0.2 fb and 0.16 fb for $\sqrt{s} = 500, 800$ GeV, respectively. As is clear in Fig. 4, the 2HDM contributions can enhance the cross section by more than one order of magnitude and reach a few fb for small m_{12} and large m_Φ. As shown in both panels of Fig. 4, we get maximum non-decoupling effect for $m_{12} = 0$. It is further amplified with large m_Φ. The plots are cut due to the perturbativity and vacuum stability constraints. The perturbativity constraints on λ_i do not allow m_Φ larger than 850 GeV. In the case of $m_{12} = 500$ GeV, the vacuum stability condition does not allow m_Φ to be less than 800 GeV. The sensitivity of cross section to $\tan\beta$ is mild.

5. To conclude, we have studied the triple Higgs couplings $\lambda_{\Phi_i\Phi_j\Phi_k}$ and the double Higgs-strahlung production $e^+e^- \to Z\Phi_i\Phi_j$ at linear collider in the framework of gen-

eral Two Higgs Doublet Models. We have quantified the sizes of the triple Higgs couplings compared to the SM triple Higgs coupling. We also show that after taking into account the perturbativity and vacuum stability constraints on the 2HDM parameters, it is possible to enhance the triple Higgs coupling $\lambda_{h^0h^0h^0}$ up to 15 times or more than the corresponding SM coupling. If the Higgs bosons $\Phi = h^0$, H^0 and A^0 are not too heavy, the double Higgs-strahlung cross sections $e^+e^- \to Z\Phi_i\Phi_j$ can be substantial, at the level of a few hundred fb, and provide some information on the triple Higgs couplings.
In the fermiophobic limit of 2HDM-I, we have pointed out that the double Higgs-strahlung processes $e^+e^- \to Zh^0h^0$ and $e^+e^- \to ZH^0h^0$ can lead to a very distinctive signature $Z2\gamma$, $Z4\gamma$, or $Z6\gamma$.
We also analyze the double Higgs-strahlung process $e^+e^- \to Zh^0h^0$ in the decoupling limit where h^0 mimics the SM Higgs boson. It is shown that in this limit, the cross section can be enhanced by about two orders of magnitude, which is much larger than the MSSM enhancement. Observations of such large cross sections would definitely indicate that the Higgs sector is 2HDM-like.

ACKNOWLEDGMENTS

AA, RB and CWC are supported by the National Science Council of R.O.C. under Grant Nos. NSC 96-2811-M-008-020, NSC 96-2811-M-033-005, and NSC 96-2112-M-008-001, respectively.

REFERENCES

1. J. F. Gunion, H. E. Haber, G. L. Kane and S. Dawson, "THE HIGGS HUNTER'S GUIDE,"(Addison–Wesley, Reading, 1990).
2. A. Djouadi, P. M. Zerwas and H. E. Haber, arXiv:hep-ph/9605437.
3. D. J. Miller and S. Moretti, Eur. Phys. J. C **13** (2000) 459.
4. A. Arhrib, R. Benbrik and C.W. Chiang arXiv:0802.0319 [hep-ph], submitted to Phys. Rev. D
5. S. Kanemura, T. Kubota and E. Takasugi, Phys. Lett. B **313**, 155 (1993). A. G. Akeroyd, A. Arhrib and E. M. Naimi, Phys. Lett. B **490**, 119 (2000).
6. T. Hahn, M. Perez-Victoria, Comput. Phys. Commun. **118**, 153 (1999); T. Hahn, Comput. Phys. Commun. **168** (2005) 78
7. J. F. Gunion and H. E. Haber, Phys. Rev. D **67**, 075019 (2003).
8. S. Kanemura, S. Kiyoura, Y. Okada, E. Senaha and C. P. Yuan, Phys. Lett. B **558** (2003) 157
9. A. Arhrib, W. Hollik, S. Penaranda and M. Capdequi Peyranere, Phys. Lett. B **579** (2004) 361.

Non-singlet sector of proton and deuteron structure functions

Ali N. Khorramian and S. Atashbar Tehrani

Physics Department, Semnan University, Semnan, Iran
and
School of Physics, IPM (Institute for Studies in Theoretical Physics and Mathematics),
P.O.Box: 19395-5531, Tehran, Iran

Abstract. We use recently calculated next-to-next-to-leading order (NNLO) anomalous dimension coefficients for the moments of the F_2 structure function in eN scattering, together with the corresponding two-loop Wilson coefficients, to obtain improved QCD predictions for non-singlet structure function. The Jacobi polynomial method is used to parameterize parton distribution up to NNLO. Target mass corrections is also consider in our calculations.

Keywords: NNLO computation, Structure function
PACS: 12.38.Bx, 14.65.-q

INTRODUCTION

Parton distribution function's (PDF's) are important ingredients in the calculation of high lepton-hadron and hadron-hadron scattering cross section. The next–to–next–to–leading order (NNLO) QCD analysis of proton and deuteron structure in the flavor non–singlet sector have been reported by the groups GRS and BBG, see [1, 2] for their recent results. Our purpose in this paper is to study the significance of the NNLO contribution and target mass corrections with using the Jacobi polynomials method. In the present paper we determine the parton momentum distribution functions $xu_v(x,Q^2)$ and $xd_v(x,Q^2)$ using the available $e(\mu)p$ and $e(\mu)d$ world data [3, 4, 5, 6, 7] up to NNLO.

The plan of the paper is to give the relevant theoretical expressions required for our analysis. Our QCD fits and results are summarized in the last section.

GENERAL FORMALISM

The combinations of parton densities for $F_2^p(x,Q^2)$ from the DIS ep process can be, up to NNLO, written as [8]

$$F_2^p(x,Q^2) = F_{2,\mathrm{NS}}^{+}(x,Q^2) + F_{2,S}(x,Q^2) + F_{2,g}(x,Q^2)\,, \tag{1}$$

with the non–singlet contribution for three active light flavors being given by

$$\frac{1}{x}F_{2,\mathrm{NS}}^{+}(x,Q^2) = \left[C_{2,q}^{(0)} + aC_{2,\mathrm{NS}}^{(1)} + a^2C_{2,\mathrm{NS}}^{(2)+}\right] \otimes \left[\frac{1}{18}q_8^+ + \frac{1}{6}q_3^+\right](x,Q^2)\,, \tag{2}$$

CP1006, *High Energy Physics and Applications*, edited by J. Ellis, S. Nasri, and E. Malkawi

here $q_3^+ = u+\bar{u}-(d+\bar{d}) = u_v - d_v$ and $q_8^+ = u+\bar{u}+d+\bar{d}-2(s+\bar{s}) = u_v+d_v+4\bar{q}-4\bar{s}$, where $\bar{q} \equiv \bar{u} = \bar{d}$ and $s=\bar{s}$. Also in the above equation $a = a(Q^2) \equiv \alpha_s(Q^2)/4\pi$ denotes the strong coupling constant and $C_i(N)$ are the Wilson coefficients [9]. The solution of the non–singlet evolution equation for the parton densities to 3–loop order is given in Ref. [2].

The flavor singlet and gluon contributions in Eq. (1) are as following

$$\frac{1}{x}F_{2,S}(x,Q^2) = \frac{2}{9}\left\{\left[C_{2,q}^{(0)} + a\,C_{2,q}^{(1)} + a^2\,C_{2,q}^{(2)}\right] \otimes \Sigma(x,Q^2)\right\}\,, \tag{3}$$

$$\frac{1}{x}F_{2,g}(x,Q^2) = \frac{2}{9}\left\{\left[a\,C_{2,g}^{(1)} + a^2\,C_{2,g}^{(2)}\right] \otimes g(x,Q^2)\right\}\,, \tag{4}$$

with $\Sigma(x,Q^2) \equiv \Sigma_{q=u,d,s}(q+\bar{q}) = u_v+d_v+4\bar{q}+2\bar{s}$. The combinations of parton densities in the valence region $x \geq 0.3$ for F_2^p in LO is

$$\frac{1}{x}F_2^p(x,Q^2) = \left[\frac{1}{18}q_{\rm NS,8}^+ + \frac{1}{6}q_{\rm NS,3}^+\right](x,Q^2) + \frac{2}{9}\Sigma(x,Q^2)\,, \tag{5}$$

where $q_{\rm NS,3}^+ = u_v - d_v$, $q_{\rm NS,8}^+ = u_v+d_v$ and $\Sigma = u_v+d_v$, since sea quarks can be neglected in the valence region $x \geq 0.3$. So in the Mellin-n space we have

$$F_2^p(n,Q^2) = \frac{5}{18}xq_{\rm NS,8}^+ + \frac{1}{6}xq_{\rm NS,3}^+ \quad = \quad \frac{4}{9}xu_v + \frac{1}{9}xd_v\,, \tag{6}$$

same of the above equation for F_2^d we have

$$F_2^d(n,Q^2) = \frac{5}{18}xq_{\rm NS,8}^+ \quad = \quad \frac{5}{18}x(u_v+d_v)\,. \tag{7}$$

In the region $x \leq 0.3$ for the difference of the proton and deuteron data we use

$$F_2^{NS}(n,Q^2) \equiv 2(F_2^p - F_2^d) = \frac{1}{3}xq_{\rm NS,3}^+ = \frac{1}{3}x(u_v-d_v) + \frac{2}{3}x(\bar{u}-\bar{d})\,, \tag{8}$$

where now $q_{\rm NS,3}^+ = u_v - d_v + 2(\bar{u}-\bar{d})$ since sea quarks cannot be neglected for x smaller than about 0.3. In our calculation we supposed the $\bar{d}-\bar{u}$ distribution from [10] at $Q_0^2 = 4$ GeV^2 which plays a marginal role in our analysis. In this work we choose the following parametrization for the valence quark densities

$$xv_i(x,Q^2) = N_i x^{a_i}(1-x)^{b_i}(1+c_i\sqrt{x}+d_i x)\,, \tag{9}$$

with $v_i = u_v, d_v$, $i = u,d$. By QCD fits of the world data for $F_2^{p,d,NS}$, we can extract valence quark densities using Jacobi polynomials method.

THE METHOD OF THE QCD ANALYSIS OF STRUCTURE FUNCTION

One of the simplest and fastest possibilities in the structure function reconstruction from the QCD predictions for its Mellin moments is Jacobi polynomials expansion. According to Ref. [11, 12, 13] we have a relation between the structure function and its Mellin moments

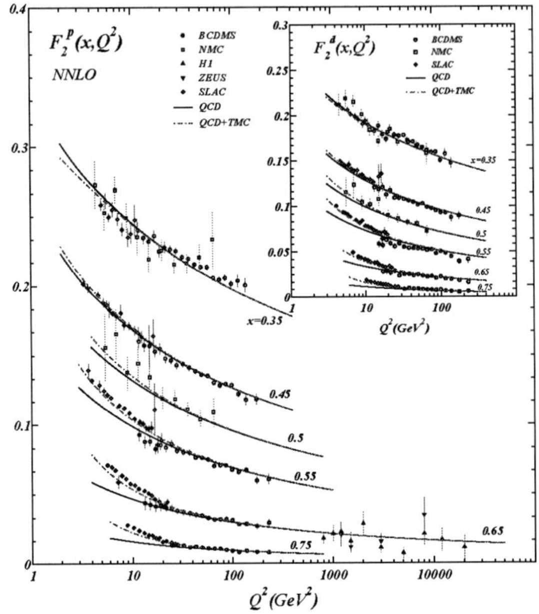

FIGURE 1. The structure function F_2^p and F_2^d as a function of Q^2 in intervals of x. Shown are the pure QCD fit in NNLO (solid line) and the contributions from target mass corrections TMC (dashed line).

$$F_2^k(x,Q^2) = x^\beta(1-x)^\alpha \sum_{n=0}^{N_{max}} \Theta_n^{\alpha,\beta}(x) \sum_{j=0}^{n} c_j^{(n)}(\alpha,\beta) F_2^k(j+2,Q^2) \,, \qquad (10)$$

where $F_2^k(j+2,Q^2)$ are the moments of $F_2^{p,d,NS}(x,Q^2)$. In our analysis we use $N_{max}=9$, $\alpha=3.0$ and $\beta=0.5$. We used all of the world data for $F_2^{p,d,NS}$ as discussed in Ref. [2]. The data in the kinematic region $Q^2 \geq 4$ GeV^2 and $4 \leq W^2 \leq 12.5$ GeV^2 may thus be used to yield information on power corrections in $F_2^p(x,Q^2)$ and $F_2^d(x,Q^2)$. For this purpose we consider the QCD fit results obtained for $W^2 \geq 12.5$ GeV^2 to the above region and from applying target mass corrections. In this paper we use the following expression [13]

$$F_2^{k(TMC)}(n,Q^2) = F_2^k(n,Q^2) + \frac{n(n+1)}{n+2}\frac{M^2_{nucl.}}{Q^2} F_2^k(n+2,Q^2) + \mathcal{O}\left(\left(\frac{M^2_{nucl.}}{Q^2}\right)^2\right) \,, \qquad (11)$$

where $M_{nucl.}$ is the mass of the nucleon, and we neglect the higher powers than $(m_N^2/Q^2)^2$. Fig.1 illustrates the proton data for $F_2(x,Q^2)$ in the valence quark region

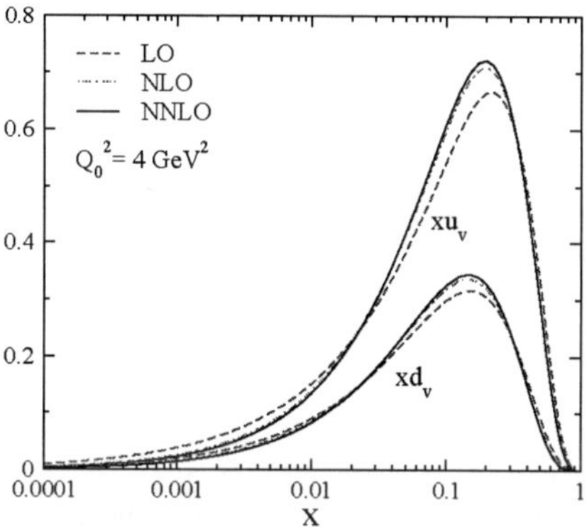

FIGURE 2. Non-singlet parton densities in LO, NLO, NNLO at $Q_0^2 = 4$ GeV^2 .

$x \geq 0.3$. The solid lines correspond to the NNLO QCD fit and the dashed lines correspond to the NNLO QCD fit adding target mass corrections. There, at higher values of x a clear gap between the data and the QCD fit is seen. Also this figure shows the corresponding results for the deuteron data. In Fig. 2 we present the valence quark densities at $Q_0^2 = 4$ GeV^2 up to NNLO. We hope to be able to take into account the higher twist effect in future, which is important at high values of x and low values of Q^2.

ACKNOWLEDGMENTS

We acknowledge Semnan university for the financial support of this project.

REFERENCES

1. M. Gluck, E. Reya and C. Schuck, Nucl. Phys. B **754** (2006) 178.
2. J. Blumlein, H. Bottcher and A. Guffanti, Nucl. Phys. B **774** (2007) 182.
3. A.C. Benvenuti *et al.* [BCDMS Collaboration], Phys. Lett. B **237** (1990) 592; A.C. Benvenuti *et al.* [BCDMS Collaboration], Phys. Lett. **B223** (1989) 485; Phys. Lett. **B237** (1990) 592. A.C. Benvenuti *et al.* [BCDMS Collaboration], Phys. Lett. B **237** (1990) 599.
4. L. W. Whitlow, E. M. Riordan, S. Dasu, S. Rock and A. Bodek, Phys. Lett. B **282** (1992).
5. M. Arneodo *et al.* [New Muon Collaboration], Nucl. Phys. B **483** (1997) 3 [arXiv:hep-ph/9610231].
6. C. Adloff *et al.* [H1 Collaboration], Eur. Phys. J. C **21** (2001) 33 [arXiv:hep-ex/0012053]; C. Adloff *et al.* [H1 Collaboration], Eur. Phys. J. C **30** (2003) 1 [arXiv:hep-ex/0304003].
7. J. Breitweg *et al.* [ZEUS Collaboration], Eur. Phys. J. C **7** (1999) 609 [arXiv:hep-ex/9809005]; S. Chekanov *et al.* [ZEUS Collaboration], Eur. Phys. J. C **21** (2001) 443 [arXiv:hep-ex/0105090].
8. M. Gluck, C. Pisano and E. Reya, Eur. Phys. J. C **50** (2007) 29 [arXiv:hep-ph/0610060].
9. J. A. M. Vermaseren, A. Vogt and S. Moch, Nucl. Phys. B **724** (2005) 3 [arXiv:hep-ph/0504242].
10. J. Blümlein, H, Böttcher, and A. Guffanti, *Nucl. Phys. B* (Proc. Suppl.) **135** (2004) 152.
11. S. Atashbar Tehrani and A. N. Khorramian, JHEP **0707** (2007) 048 [arXiv:0705.2647 [hep-ph]].
12. A. N. Khorramian, S. Atashbar Tehrani, M. Ghominejad, Acta Phys. Polon. B **38** (2007) 3551.
13. A. L. Kataev, A. V. Kotikov, G. Parente and A. V. Sidorov, Phys. Lett. B **417** (1998) 374 [arXiv:hep-ph/9706534].

Universal Extra Dimension models with right-handed neutrinos

Shigeki Matsumoto*, Joe Sato†, Masato Senami** and Masato Yamanaka†

*Institute for International Advanced Interdisciplinary Research, Tohoku University, Sendai, Miyagi 980-8578, Japan
†Department of Physics, Saitama University, Shimo-okubo, Sakura-ku, Saitama, 338-8570, Japan
**ICRR, University of Tokyo, Kashiwa, Chiba 277-8582, Japan

Abstract. Relic abundance of dark matter is investigated in the framework of universal extra dimension (UED) models with right-handed neutrinos. These models are free from the KK graviton problem in the minimal UED model. The first KK particle of the right-handed neutrino is a dark matter candidate in this framework. When ordinary neutrino masses are large enough such as the degenerate mass spectrum case, the dark matter relic abundance can increase significantly. The scale of the extra dimension consistent with cosmological observations can be 500 GeV in the minimal setup of UED models with right-handed neutrinos.

Keywords: dark matter, extra dimension, neutrino
PACS: 95.35.+d

INTRODUCTION AND UED MODELS

There is dark matter in the universe, and many models beyond the Standard Model (SM) have been proposed to explain the dark matter. Among those, Universal Extra Dimension (UED) models are one of interesting candidates for new physics, and it is worth investigating these models. We have solved problems inherent in these models, and calculated the allowed parameter by estimating the dark matter relic abundance. This proceeding is based on our works [1, 2].

First, we briefly review UED model. This model is described in the five-dimensional space-time, where the extra dimension is compactified on an S^1/Z_2 orbifold with the radius R. As a result of the compactification, many excited states of SM fields, called KK particles, appear. In this model, the lightest KK particle (LKP) is stabilized by discrete symmetry, called KK parity. Thus, if the LKP is neutral, the LKP can be dark matter candidate.

Though the minimal UED model is good model, the model has two shortcomings. The first one is the KK graviton problem. In the parameter region where $1/R < 800$ GeV, the KK graviton $G^{(1)}$ is the LKP, and the next LKP (NLKP) is the KK photon $\gamma^{(1)}$. Hence, $\gamma^{(1)}$ produced in the early universe decay into photons at late time universe, and these photons distort the CMB spectrum or the diffuse photon spectrum.

The second problem is the absence of neutrino masses. Since UED model has been constructed as minimal extension of the SM, neutrinos are treated as massless particles. However, neutrinos have mass. Therefore we must introduce neutrino masses into UED models.

CP1006, *High Energy Physics and Applications*, edited by J. Ellis, S. Nasri, and E. Malkawi

SOLVING THE PROBLEMS

In order to solve these problems, we introduce the right-handed neutrinos into UED models, and assume that they form Dirac mass with ordinary neutrinos. The neutrino masses are expressed as $\mathscr{L}_\nu = y_\nu \bar{N} L \Phi + \text{h.c.}$, where N is the right-handed neutrino, L is the left-handed lepton. Then the second problem, the absence of the neutrino masses, are clearly solved. Once we introduce right-handed neutrinos into UED models, their KK particles automatically appear in the spectrum. The mass of the first KK right-handed neutrino, $N^{(1)}$, is estimated as $m_{N^{(1)}} \simeq \frac{1}{R} + O\left(\frac{m_\nu^2}{1/R}\right)$. After introducing the right-handed neutrinos, $N^{(1)}$ is the NLKP, and $\gamma^{(1)}$ is the next to next lightest KK particle.

The existence of the $N^{(1)}$ NLKP changes the late time decay of $\gamma^{(1)}$. In the models with the right-handed neutrino, $\gamma^{(1)}$ dominantly decays into $N^{(1)}$ and SM left-handed neutrino at tree level ($\gamma^{(1)} \to N^{(1)}\bar{\nu}$). In the KK neutral gauge boson mass matrix, mixing angle is very small, and hence $\gamma^{(1)}$ can be regarded as KK U(1) gauge boson. Therefore $\gamma^{(1)}$ can decay into neutrinos through the hypercharge. On the other hand, the dominant decay mode associated with a photon is $\gamma^{(1)} \to G^{(1)}\gamma$. We estimeted the branching ratio of these decay modes

$$\text{Br} = \frac{\Gamma(\gamma^{(1)} \to G^{(1)}\gamma)}{\Gamma(\gamma^{(1)} \to N^{(1)}\bar{\nu})} = 5 \times 10^{-7} \; \frac{1/R}{500\text{GeV}}{}^{3} \; \frac{0.1\text{eV}}{m_\nu}{}^{2} \; \frac{\delta m}{1\text{GeV}} \; . \tag{1}$$

As a result, by introducing the right-handed neutrino into UED models, neutrino masses are introduced, and problematic high energy photon emission is highly suppressed. Therefore, two problems in UED models have been solved simultaneously.

$N^{(1)}$ DARK MATTER

$N^{(1)}$ cannot decay because it is forbidden by the kinematics. Since $N^{(1)}$ is neutral, massive, and stable, $N^{(1)}$ is also a dark matter candidate by introducing the right-handed neutrinos. Though abundance produced from decoupled $\gamma^{(1)}$ decay is same, abundance of $N^{(1)}$ dark matter produced from thermal bath is larger than that of $G^{(1)}$. Therefore, when $N^{(1)}$ is dominant component of dark matter, there are additional contribution to the dark matter relic abundance. As total dark matter number density becomes large, the dark matter mass allowed by cosmological observations, $\sim 1/R$, becomes small. Therefore, in order to determine the compactification scale $1/R$, we must evaluate the number density of $N^{(1)}$ dark matter.

Because the number density of the decoupled $\gamma^{(1)}$ has been calculated in previous works [3], we know $N^{(1)}$ number density produced from the decay of that. So, we calculate the $N^{(1)}$ number density produced from the thermal bath. To do this, we must include the thermal effects. In the thermal bath, particle mass receives thermal corrections. For example, the KK Higgs boson $\Phi^{(n)}$ mass is,

$$m^2_{\Phi^{(n)}}(T) = m^2_{\Phi^{(n)}}(T=0) + \; a(T)\cdot 3\lambda_h + x(T)\cdot 3y_t^2 \; \frac{T^2}{12} \; . \tag{2}$$

Coefficients $a(T)$ and $x(T)$ are counting factor how many KK modes contribute to the correction at the temperature T. Taking account of thermal masses, we calculated the $N^{(1)}$ number density.

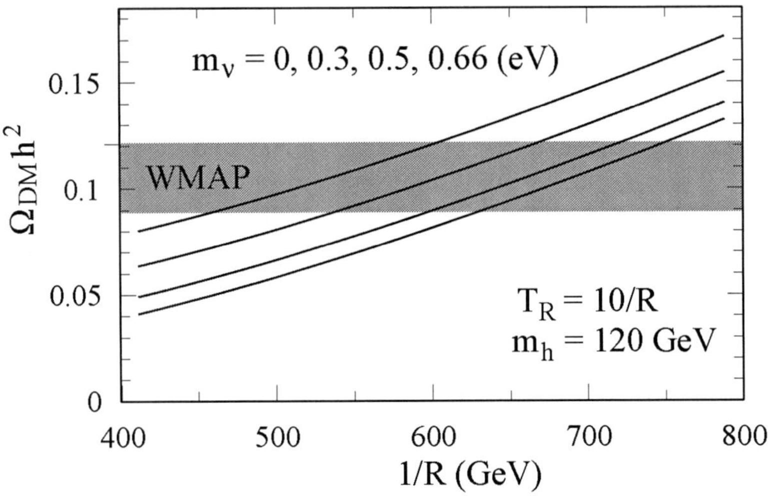

FIGURE 1. The dependence of the abundance on m_ν with fixed reheating temperature $T_R = 10/R$ and $m_h = 120$ GeV. The solid lines are the relic abundance for $m_\nu = 0, 0.3, 0.5, 0.66$ eV from bottom to top. The gray band represents the allowed region from the WMAP observation at the 2σ level

Finally, we show the numerical result. Figure 1 shows the neutrino mass dependence of the abundance. The horizontal axis is the compactification scale $1/R$, and the vertical axis is the dark matter relic abundance. The lines correspond to the result with $m_\nu = 0, 0.3, 0.5, 0.66$eV from bottom to top. As shown in this figure, for large m_ν, compactification scale can be less than 500 GeV. If compactification scale is less than 500 GeV, in ILC experiment, n=2 KK particles can be produced. n=2 KK particles are very important for discriminating UED from SUSY at collider experiment.

SUMMARY

We have solved two problems in UED models (absence of the neutrino mass, KK graviton problem) by introducing the right-handed neutrinos. We have shown that by introducing right-handed neutrino, the dark matter is the KK right-handed neutrino $N^{(1)}$, and we have calculated the relic abundance of the $N^{(1)}$ dark matter. In our model, the compactification scale $1/R$ can be less than 500 GeV. This fact has important consequence on the collider physics.

REFERENCES

1. S. Matsumoto, J. Sato, M. Senami and M. Yamanaka, *Phys. Rev.* D **76**, 043528 (2007)

2. S. Matsumoto, J. Sato, M. Senami and M. Yamanaka, *Phys. Lett.* B **647**, 466 (2007)
3. M. Kakizaki, S. Matsumoto and M. Senami, *Phys. Rev.* D **74** (2006) 023504;
   S. Matsumoto and M. Senami, *Phys. Lett.* B **633** (2006) 671

A solution to the ^{7}Li problem by the long lived stau

Toshifumi Jittoh*, Kazunori Kohri†, Masafumi Koike*, Joe Sato*, Takashi Shimomura* and Masato Yamanaka*

*Department of Physics, Saitama University, Shimo-Okubo, Sakura-ku, Saitama, 338-8570, Japan
†Physics Department, Lancaster University LA1 4YB, UK

Abstract. Modification of standard big-bang nucleosynthesis is considered in the minimal supersymmetric standard model to resolve the excessive theoretical prediction of the abundance of primordial lithium 7. This talk is based on [1] was given by J. Sato.

Keywords: long-lived stau, ^{7}Li problem, internal conversion
PACS: 12.60.Jv, 14.80.Ly, 26.35.+c, 98.80.Cq

INTRODUCTION

The theory of big-bang nucleosynthesis (BBN) has been successful in predicting the abundance of light elements in the Universe from baryon-to-photon ratio η. The recent results of the Wilkinson microwave anisotropy probe (WMAP) experiment [2], however, put this theory into challenge. The standard BBN (SBBN) predicts ^{7}Li to be $(4.15^{+0.49}_{-0.45}) \times 10^{-10}$[3]. This prediction is inconsistent with the observation of metal-poor halo stars which implies $(1.23^{+0.32}_{-0.25}) \times 10^{-10}$ [4].

In this paper we investigate a role of the bound state of exotic charged massive particles (CHAMPs) and a nucleus. We put the CHAMP BBN scenario in the minimal supersymmetric standard model (MSSM) with the conservation of *R* parity. A possible candidate of CHAMPs is the next-lightest superparticle (NLSP) with electric charge, which can have a long lifetime by assuming a small mass difference from the LSP [5].

We assume in the present paper that the LSP is a neutralino $\tilde{\chi}_0$ and the NLSP is a stau $\tilde{\tau}$, the superpartner of tau lepton τ. We consider the bound state of ^{7}Be and $\tilde{\tau}^-$ in the early Universe and the subsequent decay chain of nucleus $^7\mathrm{Be} \to {}^7\mathrm{Li} \to {}^7\mathrm{He}$ due to the interactions of the two. The ^{7}He nuclei rapidly decay into ^{6}He nuclei, which are effectively stable in the considered time scale. With the freedom of the mass of stau $m_{\tilde{\tau}}$ and its lifetime $\tau_{\tilde{\tau}}$, we search for the possible solution to the ^{7}Li problem that are phenomenologically acceptable.

INTERACTIONS OF STAUS WITH ^{7}BE AND ^{7}LI

In this section, we consider the most inportant stau-nucleus interaction processe, internal conversion of stau-nucleus bound state.

CP1006, *High Energy Physics and Applications*, edited by J. Ellis, S. Nasri, and E. Malkawi

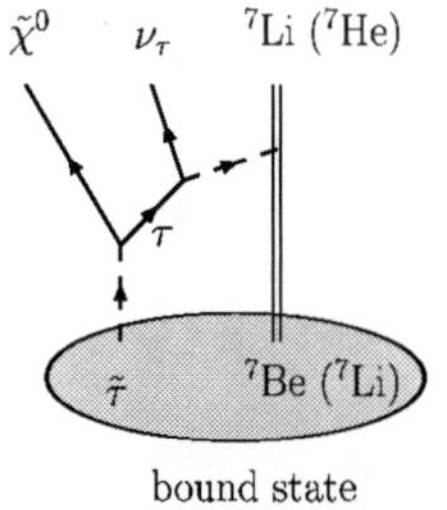

FIGURE 1. The Feynmann diagrams of internal conversion of ^{7}Be (^{7}Li).

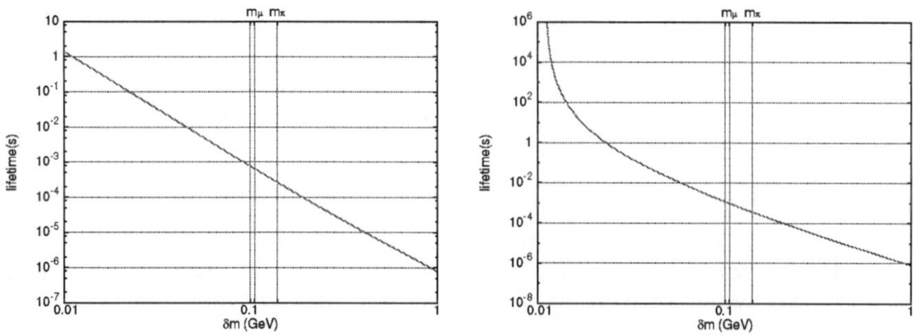

FIGURE 2. (color online). The lifetimes of internal conversion processes as the function of δm. Top panel: $(\tilde{\tau}^7\text{Be}) \rightarrow \tilde{\chi}^0 + \nu_\tau + {}^7\text{Li}$, bottom panel: $(\tilde{\tau}^7\text{Li}) \rightarrow \tilde{\chi}^0 + \nu_\tau + {}^7\text{He}$. We take $m_{\tilde{\chi}^0} = 300\text{GeV}$. Other parameters are shown in [1]

The interaction between a stau and a nucleus proceeds more efficiently when they form a bound state (see Fig. 1) [6] due to two reasons: (1) the overlap of the wave functions of the two becomes large since the stau and particle are packed in the small space, (2) the small distance between the two allows virtual exchange of the hadronic current even if $\delta m < m_\pi$. The stau-nucleus bound state decays through the following processes:

$$\tilde{\tau} + {}^7\text{Be} \rightarrow (\tilde{\tau}^7\text{Be}) \rightarrow \tilde{\chi}^0 + \nu_\tau + {}^7\text{Li}, \tag{1}$$

$$\tilde{\tau} + {}^7\text{Li} \rightarrow (\tilde{\tau}^7\text{Li}) \rightarrow \tilde{\chi}^0 + \nu_\tau + {}^7\text{He}, \tag{2}$$

$${}^7\text{He} \rightarrow {}^6\text{He} + \text{n}, \tag{3}$$

$${}^6\text{He} + \text{background particles} \rightarrow {}^3\text{He}, {}^4\text{He}, \text{etc.}, \tag{4}$$

where the parentheses denote the bound states. We note that we introduce not only reaction (1), but also reaction (2). The ^{6}He nucleus can also decay into ^{6}Li via β decay with the lifetime 817 msec. We do not take this process into account since this process is much slower than the scattering process (4).

The lifetime of the internal conversion τ_{IC} are shown in Fig.2 as a function of the mass difference. We find that the lifetime of the internal conversion process is in the order of

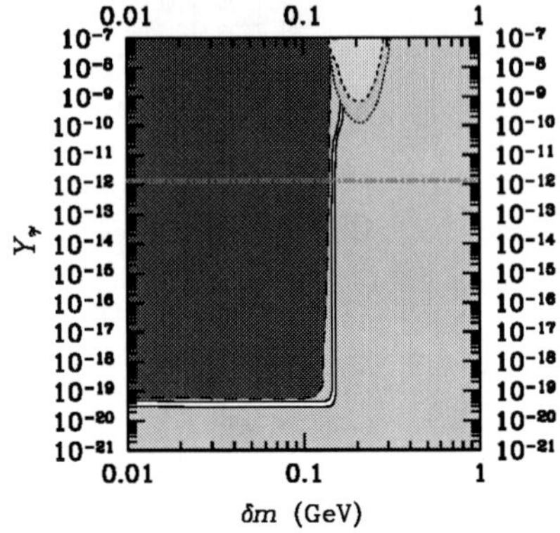

FIGURE 3. (color online). The constraints from the light-element abundance shown in the δm–$Y_{\tilde{\tau}}$ plane.

10^{-3} sec. They are short enough so that this internal conversion processes contribute the destruction of ^{7}Be (^{7}Li) significantly. The internal conversion processes are indeed dominant over the other processes.

NUMERICAL CALCULATION AND INTERPRETATION OF THE RESULT

In this section, we study the effectiveness of new decay channels on the ^{7}Li problem by numerical calculation. To do this, we choose the abundance of stau $Y_{\tilde{\tau}} \equiv n_{\tilde{\tau}}/s|_{t=t_{\text{freeze out}}}$ at freeze out time and mass difference δm as free parameters.

In Fig. 3 we show the parameter region of δm and $Y_{\tilde{\tau}}$ allowed by the observational light-element abundances. The white region is consistent with all the observational abundance including ^{7}Li to hydrogen ratio (^{7}Li/H). The regions enclosed by dotted (green), dashed (light blue), and dash-dotted (purple) lines are excluded by the observations on the mass fraction of ^{4}He (Y_p), deuterium to hydrogen ratio (D/H) and ^{6}Li to ^{7}Li ratio (^{6}Li/^{7}Li), respectively. We adopt the following observational constraints on primordial light element abundances

$$
\begin{aligned}
Y_\text{p}^{\text{obs}} &= 0.2516 \pm 0.0040 \text{ [7]}, & (5)\\
(\text{D/H})^{\text{obs}} &= (2.82 \pm 0.26) \times 10^{-5} \text{ [8]}, & (6)\\
\text{Log}_{10}\left({}^7\text{Li/H}\right)^{\text{obs}} &= -9.91 \pm 0.10 \text{ [4]}, & (7)\\
\left({}^6\text{Li}/{}^7\text{Li}\right)^{\text{obs}} &< 0.046 \pm 0.022 + 0.84 \text{ [9]}. & (8)
\end{aligned}
$$

The qualitative feature of the allowed region is expalained from the following physical consideration. First, staus need to have lifetimes $\tau_{\tilde{\tau}}$ longer than the time required to form the bound state of a stau and a ^{7}Be. The required time $t_{\text{form},^7\text{Be}}$ is estimated from the binding energy $E_{\mathrm{bin},^7\mathrm{Be}} = 1490\mathrm{keV}$ as $t_{\mathrm{form},^7\mathrm{Be}} \sim 10^9\mathrm{sec}\cdot(E_{\mathrm{bin},^7\mathrm{Be}}/\mathrm{kev})^{-2} \sim \mathcal{O}(10^2\mathrm{sec})$. Since $\delta m \lesssim (100-200)\mathrm{MeV}$ is required for $\tau_{\tilde{\tau}} \gtrsim t_{\mathrm{form},^7\mathrm{Be}}$, the allowed region appears only in this region. Second, the yield value of staus $Y_{\tilde{\tau}}$ needs to be large compared with that of ^{7}Li which we denote by $Y_{^7\mathrm{Li}}$. We estimate $Y_{^7\mathrm{Li}}$ from the hydrogen to entropy ratio $n_\mathrm{H}/s \sim \mathcal{O}(10^{-10})$ and the constraint (7) as

$$Y_{^7\mathrm{Li}} \sim \left(\frac{n_{^7\mathrm{Li}}}{n_\mathrm{H}}\right)^{\mathrm{obs}} \cdot \frac{n_\mathrm{H}}{s} \sim 10^{-9.63} \times 10^{-10} \sim 10^{-20}. \tag{9}$$

We thus need $Y_{\tilde{\tau}} > 10^{-20}$, and again the allowed region appears in this region. Third, the excessive destruction of ^{7}Li by the process (2) needs to be avoided due to the constraint (8). This condition puts a limit to the formation rate of the bound state of a stau and a ^{7}Li. We then need either $\tau_{\tilde{\tau}} < t_{\mathrm{form},^7\mathrm{Li}} \sim 10^9\mathrm{sec}\cdot(E_{\mathrm{bin},^7\mathrm{Li}}/\mathrm{kev})^{-2} \sim \mathcal{O}(10^3\mathrm{sec})$, or $Y_{\tilde{\tau}}$ to be small enough. The former condition leads to $\delta m \gtrsim 100\mathrm{MeV}$, although this region is subject to the strong restriction considered in the previous paragraph. The latter imposes an upper limit on $Y_{\tilde{\tau}}$ in the region $\delta m \lesssim 100\mathrm{MeV}$, and Fig. 3 suggests that this limit is in the order of 10^{-20}. The tininess of $Y_{\tilde{\tau}}$ shows that the internal conversion processes (1) and (2) are dominant over other processes such as the pion exchange and the stau-catalyzed fusion. We can confirm this dominance by an explicit calculation of the rates of these processes.

SUMMARY

We have investigated a possible solution of the ^{7}Li problem in a framework of MSSM, in which the LSP and the NLSP are neutralino and stau, respectively, and have a tiny mass difference of $\delta m \lesssim (100-200)\mathrm{MeV}$. The staus then survive throughout the BBN era and provide additional decay processes as mentioned in Sec. to reduce the primordial ^{7}Li abundance.

Taking the three new processes into account, we numerically calculated the primordial abundance of light elements varying the LSP-NLSP mass difference and the abundance of stau. Taking $\theta_\tau = \pi/3,\ \gamma_\tau = 0$, we obtained the parameter region consistent to the observed ^{7}Li abundance.

REFERENCES

1. Toshifumi Jittoh, Kazunori Kohri, Masafumi Koike, Joe Sato, Takashi Shimomura, and Masato Yamanaka Phys. Rev. **D76**, 125023, (2007)
2. D. N. Spergel et al., Astrophys. J. Suppl. Ser. **170**, 377 (2007)
3. A. Coc, E. Vangioni-Flam, P. Descouvemont, A. Adahchour and C. Angulo, Astrophys. J. **600**, 544 (2004).
4. S. G. Ryan, T. C. Beers, K. A. Olive, B. D. Fields and J. E. Norris, Astrophys. J. Lett. **530**, L57 (2000).
5. T. Jittoh, J. Sato, T. Shimomura and M. Yamanaka, Phys. Rev. D **73**, 055009 (2006).
6. C. Bird, K. Koopmans and M. Pospelov, arXiv:hep-ph/0703096.
7. Y. I. Izotov, T. X. Thuan and G. Stasinska, arXiv:astro-ph/0702072.
8. J. M. O'Meara, S. Burles, J. X. Prochaska, G. E. Prochter, R. A. Bernstein and K. M. Burgess, Astrophys. J. **649**, L61 (2006).
9. M. Asplund, D. L. Lambert, P. E. Nissen, F. Primas and V. V. Smith, Astrophys. J. **644**, 229 (2006).

Beyond the Standard Model: Non Commutative Geometry Approach and LHC Signatures

N.Mebarki[a]

[a] *Laboratoire de Physique Mathématique et Subatomique, Faculty of Science, Mentouri University, Constantine, Algeria*

Abstract. Some noncommutative space-time standard model features and possible LHC signatures are presented. The spin effects on the modified noncommutative Altarelli-Parisi splitting functions for polarized partons are also discussed.

Keywords: Noncommutative geometry, Beyond standard model, Polarization.
PACS: 11.10.Nx, 13.88.+e, 12.60.-i

INTRODUCTION

Over the past recent years, the thrust of "beyond the standard model" theory has undergone an important progress, from model extensions to space-time structure modifications namely the noncommutativity[1]-[7]. The simplest formulation of a noncommutative geometry (NCG) standard model maintaining a "deformed" star-* gauge invariance was realized using a nonlinear field redefinition that is determined order by order in an expansion in the noncommutative parameter ($\theta_{\mu\nu},\ \left|\theta_{\mu\nu}\right| \approx 1/\Lambda_{NC}^2$) and called the Seiberg-Witten maps [1].

NCG FEATURES AND LHC SIGNATURES

At the LHC, one needs a theoretical control of partons distribution functions (PDFs) shapes,errors and look for tools for new physics. Noncommutative geometry as a possible alternative opens new windows, search for specific scenarios and signatures (missing energy, spin correlations etc...). In what follows, we use the mathematical formalism of ref.(1) and the non minimal noncommutative standard model Feynmann rules of ref.(8) where the strong interactions are also included. It is worth to mention that in this approach, and apart from the non-commutative corrections, several new interactions appear. The most interesting ones are gluonic interactions with the electroweak sector. Regarding, the current bounds on Λ_{NC} (if we assume a Lorentz violation where $\theta_{\mu\nu}$ are constant), they are strongly model dependent, ranging from 141 GeV in collider experiments, 2.5 TeV for dilepton production in $\gamma\gamma$ collisions and up to 10^{14} GeV from rotation invariance tests in atomic physics and astrophysics[9], [10]. One can have also a Lorentz non violation if $\theta_{\mu\nu}$ is a tensor [11].

CP1006, *High Energy Physics and Applications*, edited by J. Ellis, S. Nasri, and E. Malkawi

NCG Polarized Altarelli-Parisi Splitting Functions

Transverse polarization distribution and fragmentation functions parameterize transverse spin effects in hard scattering processes. Although in general little is known experimentally on most of these functions, some transverse spin experiments have been performed and large single transverse spin asymmetries have been observed [12] in the process $pp\uparrow \rightarrow \pi X$. The main observation is that the description of this specific process in terms of transverse spin functions will lead to power suppressed single spin asymmetries (SSA), unless one takes into account the intrinsic transverse momentum (K_T) of the partons. As it will be clear later, we will see that the space-time noncommutativity will play similar role as the K_T effect does. For a qualitative study, we will follow the simplest method of Altarelli-Parisi approach [13] within the non minimal NCG standard model [8]. Now, imposing parity conservation implies that:

$$P_{A_+B_\pm}(z) = P_{A_-B_\mp}(z) \tag{1}$$

Where $P_{AB}(z)$ means the probability of finding a parton A inside a parton B with a fraction z of the longitudinal momentum of A ('+' and '-.' refer to positive and negative helicities). If masses are neglected, the vector quark-gluon interaction is helicity conserving and therefore:

$$P_{q_-q_+}(z) = P_{q_+q_-}(z) = 0 \tag{2}$$

Furthermore, using the fact that the total number of quarks minus antiquarks has to be conserved (charges and current algebra sum rules are protected against radiative corrections) and the momentum of the proton is unchanged, one has to have the following constraints:

$$\int_0^1 dzz(2N_f P_{q_+G_\pm}(z) + P_{G_+G_\pm}(z)) = 0, \quad \int_0^1 dzz(P_{q_+q_\pm}(z) + P_{G_+q_\pm}(z)) = 0, \quad \int_0^1 dzzP_{q_+q_+}(z) \tag{3}$$

Now, the new noncommutative quark-quark-gluon and three gluon vertices[8] together with the polarization four vector for positive and negative helecities lead to:

$$P_{q_+q_+}(z) = P_{q_-q_-}(z) = C_2(R)\left[\frac{1+z^2+x_T^2/2}{2z(1-z)_+} + \frac{3}{2}\delta(1-z) + \frac{1}{2}x_T^2(\frac{1}{z_+} + \frac{1}{(z^2)_{++}})\right] \tag{4}$$

$$P_{G_+q_+}(z) = C_2(R)(\frac{1}{z} + \frac{x_T^2}{2z((1-z)^2)_{++}} - \frac{x_T^2}{2}\delta(1-z)), \quad P_{G_-q_+}(z) = C_2(R)\frac{1}{z}((1-z)^2 + \frac{x_T^2}{2}) \tag{5}$$

$$P_{q+G_+}(z) = \frac{1}{4}\left[z^2 + x_T^2\frac{(2z-1)^2}{8z}(\frac{z}{((1-z)^2)_{++}} + 8B\delta(1-z))\right], P_{q_-G_+}(z) = \frac{1}{4}\left[(1-z)^2 + x_T^2\frac{(2z-1)^2}{zz_+}\right] \tag{6}$$

$$P_{G_+G_+}(z) = C_2(G)((1+z^4) + (\frac{1}{z} + \frac{1}{(1-z)_+}) + (\frac{11}{6} - \frac{2T(R)}{3C_2(G)})\delta(1-z) + x_T^2 f_{G_+G_+}) \tag{7}$$

$$P_{G_-G_+}(z) = C_2(G)\left[\frac{(1-z)^3}{z} + x_T^2\frac{(1+z^2+(1-z)^2)^2}{z}(\frac{g_2}{(z^2)_{++}})\right] \tag{8}$$

with

$$f_{G_+G_+} = \frac{(1+z^2+(1-z)^2)^2}{z}(\frac{g_1}{(z^2)_{++}} + \frac{2}{(1-z)_+} + \frac{1}{((1-z)^3)_{+++}} + \frac{A}{4}\delta(1-z)) \tag{9}$$

and

$$B+2\frac{C_2(G)}{N_f}A=\frac{13}{6}+\frac{46}{3}\frac{C_2(G)}{N_f},\quad g_1+g_2=1+z \tag{10}$$

where the Casimir operators $C_2(R)$, $C_2(G)$ and $T(R)$ are given by

$$C_2(R)=(N_c^2-1)/N_c,\quad C_2(G)=N_c,\quad T(R)=N_f/2 \tag{11}$$

(N_c and N_f are the color and flavor numbers respectively). Notice the ambiguities in defining the kernels $P_{G_+G_+}$, $P_{G_-G_+}$ and $P_{q_+G_+}$. The distributions $1/(1-z)_+$, $1/(1-z)_{++}$ and $1/(1-z)_{+++}$ are such that:

$$\int_0^1 dz\frac{f(z)}{(1-z)_+}=\int_0^1 dz\frac{f(z)-f(1)}{(1-z)},\quad \int_0^1 dz\frac{f(z)}{((1-z)^2)_{++}}=\int_0^1 dz\frac{f(z)-f(1)+(1-z)f'(1)}{(1-z)^2} \tag{12}$$

and

$$\int_0^1 dz\frac{f(z)}{((1-z)^3)_{+++}}=\int_0^1 dz\frac{f(z)-f(1)+(1-z)f'(1)-(1-z)^2 f''(1)/2}{(1-z)^3} \tag{13}$$

Similar definitions can be set for the distributions $1/z_+$ and $1/z_{++}$. Here $x_T=P_T/\Lambda_{NC}$ where p_T is some intrinsic small transverse momentum of the radiated parton. To simplify matter, we take only the non vanishing component $\theta^{01}=\eta=1/\Lambda_{NC}^2$. Notice that, in the expressions of the polarized splitting functions, the terms proportional to x_T^2 are more singular than those calculated within the classical commutative QCD and this will affect the (DGLAP) evolution equation especially at smaller partons momemtum fraction x. Notice also, that the noncommutativity is related to the intrinsic transverse motion of the partons which is crucial for the spin structure studies leading to SSA and 3D dimensional description of the pdfs in terms of (x, Q^2, P_T/Λ_{NC}).

NCG Diphotons Production

The perturbative contribution to $pp\to\gamma\gamma+X$ proceeds at the lowest order g^2 (g is the electromagnetic coupling constant) via the quark annihilation subprocess $q\bar{q}\to\gamma\gamma$. The remaining important perturbative contribution is the gluon annihilation subprocess $gg\to\gamma\gamma$. Although this arises at order $g^2g_s^2$ (g_s is the strong coupling) through a quark box loop, it is still of the same size as the quark annihilation subprocess, due to the large gluon luminosity at the LHC. In NCG, and in addition to the contribution of the

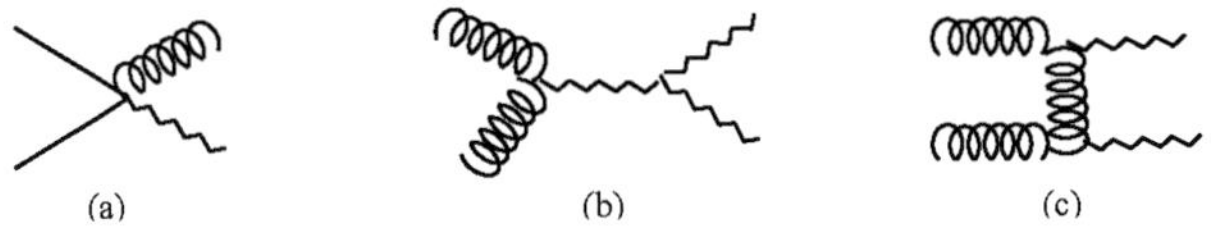

FIGURE.1.Some pure NCG Feynman diagrams for:(a) $q\bar{q}\to\gamma\gamma$,(b),(c) $gg\to\gamma\gamma$

modified vertices in the above mentioned subprocesses, we have new pure NCG Feynman diagrams (see figs.(1)) proportional to $\eta g g_s s$ (fig.1a) and $\eta^2 g^2 s^2$ (fig.1b,c) because of factors like $k_\mu\theta^{\mu\nu}p_\nu$ in the vertices (k and p are the four momenta of the incoming and outgoing particles). Here $\sqrt{s}$ is the center of mass energy. Now, if η is of $O(TeV)$, these diagrams become important and at least comparable to higher order corrections (HOC) especially those of (fig.1b,c) where the pdf of the gluon is very important. Now, to get an idea about the new NCG features,and for a sake of illustration, we have considered just the contribution of the subprocess $q\bar{q} \to \gamma\gamma$ to the ratio $R = \hat{\sigma}_{NCG} / \hat{\sigma}_{CG}$ (see fig.2a) where $\hat{\sigma}_{CG}$ and $\hat{\sigma}_{NCG}$ are the Born commutative and noncommutative differential cross sections of the physical process $pp \to \gamma\gamma + X$ as a function of the collision angle θ and for various values of the noncommutative scale Λ_{NC}. For a space-space noncommutativity where the only non vanishing component is θ^{12} and for a fixed value of Λ_{NC} the ratio R is an increasing function of the cosine of the collision angle θ denoted by cos(theta)(see fig.2a),for example for $\Lambda_{NC} \approx 14{,}1 TeV$ and if $\cos\theta \approx 0.15$ we have $R \approx 0.88$;however,if $\cos\theta \approx 0.25$, $R \approx 0{,}93$ .Notice that for $\cos\theta \leq 0.35$, one has $R \leq 1$ (NCG contribution with negative sign)and for $\cos\theta \geq 0.35$, $R \geq 1$ .Now, for a fixed value of θ, and for small values of $\cos\theta$,the ratio R is approximately an increasing function of Λ_{NC} .However, for large values, R is a decreasing function. For example for $\cos\theta \approx 0.5$ and if $\Lambda_{NC} \approx 14{,}1 TeV$, $R \approx 1.16$; however, if $\Lambda_{NC} \approx 18{,}3 TeV$, $R \approx 1.09$. Now, for $\cos\theta \approx 0.01$ and if $\Lambda_{NC} \approx 18{,}3 TeV$ and $\Lambda_{NC} \approx 15{,}8 TeV$, R takes the values 0,89 and 0,92 respectively. To see another feature of NCG, we have considered the left-right asymmetry $A_{L,R} = \dfrac{R(\cos\theta) - R(-\cos\theta)}{R(\cos\theta) + R(-\cos\theta)}$ for a space-time noncommutativity where the only non vanishing component is θ^{01} (see fig.2b). For a fixed value of Λ_{NC} (resp. θ), $A_{L,R}$ is a decreasing function of θ (resp. Λ_{NC}). For example, for $\Lambda_{NC} \approx 14{,}1 TeV$ and if $\cos\theta$ changes from 0.15 to 0.5 , $A_{L,R}$ increases from 0.5 to 0.7 Similarly, for $\cos\theta \approx 0.75$, $A_{L,R}$ decreases from 0.59 to 0.33 if Λ_{NC} changes from $15{,}8 TeV$ to $20 TeV$.

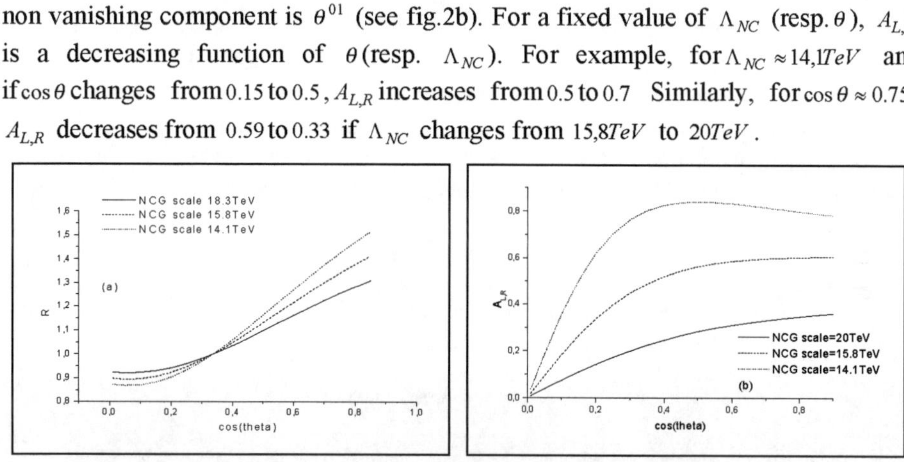

FIGURE.2. For various values of Λ_{NC} (a) Ratio R versus $\cos\theta$ (b) $A_{L,R}$ versus $\cos\theta$

NCG Associated Higgs Production

The Higgs production with an associated Z boson in $pp \to HZ$ where besides the contribution of the modified vertices, there are also important new pure NCG diagrams like the one in fig.3d which is proportional to $\eta g^2 s$. We remind also that although the box or triangle diagrams of figs.3b and 3c respectively are of higher order ($\propto g^2 g_s^2$), they are as important as the diagram of fig.3a. This is due essentially to the gluon luminosity. Now, if Λ_{NC} is of $O(TeV)$ and because of the importance of the gluon pdf, then the diagram of fig.3d will be also comparable to the other contributions ($\propto \eta gs$) or at least to the ones of HOC. Thus, we will have an enhancement in the cross section of the process $pp \to HZ$ at the LHC where the effect of the noncommutativity becomes more or less important. Moreover and because of the vertices proportional to $k_\mu \theta^{\mu\nu} p_\nu$ and $\theta^{\mu\nu} k_\nu$ in NCG diagrams involving the collision angle θ, we expect a change in the angular distribution.

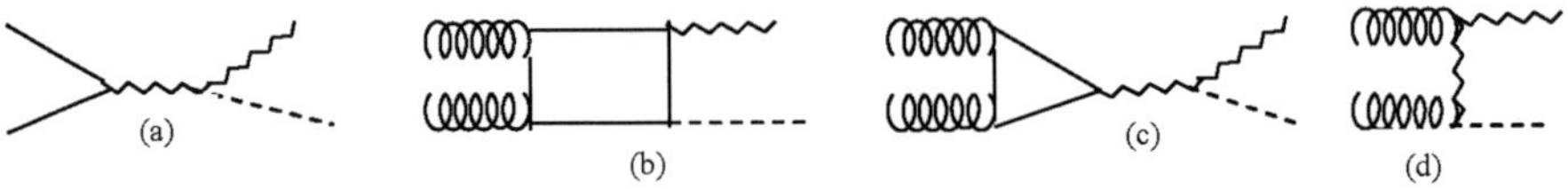

FIGURE.3.Feynman diagrams for: (a) $q\bar{q} \to HZ$, $gg \to HZ$ ((b)box ,(c) triangle and (d) pure NCG)

CONCLUSION

From our qualitative study of splitting functions, diphoton and associated higgs boson production within the space-time NCG, we expect new features namely:

a) Modification of the splitting functions with more singular terms $\propto x_T^2$ and this will affect the Q^2 evolution of the pdfs and the DGLAP equation. Moreover, we have noticed also the great similarity with the intrinsic transverse momentum.

b) Change in the shape and angular distribution of the ratio R and sensitivity to Λ_{NC}.

c) Existence of a left-right asymmetry $A_{L,R}$ in diphoton production.

d) Possible enhancement in the cross sections like in the associated higgs boson case

e)Strong correlation with the spin and beams polarization.

ACKNOWLEDGMENTS

I am very grateful to the organizers of the UAE-CERN Workshop and the Algerian Ministry of education and research for the financial support.

REFERENCES

1. X.Calmet, B.Jurco, P.Schupp, J.Wess and M.Wohlgenamt, *Euro. Phys.J.C*23, 363-376 (2002).
2. B.Jurco, L.Moller, S.Schraml, P.Schupp and J.Wess, *Euro. Phys.J.C*21, 383-388 (2001).
3. I.Hinchliffe, N.Kersting and Y.L.Ma, *Int.Journ.of Mod. Phys. A*19, 179-204 (2004).
4. P.Aschieri, B.Jurco, P.Schupp and J.Wess, *Nucl.Phys.B*651, 45-70 (2003).
5. W.Behr,N.G.Deshpande,G.Duplancic,P.Schupp,J.Trampetic,J.Wess,Euro*Phys.J.C*29,441-446(2003)
6. B P.Schupp, J.Trampetic, J.Wess and G.Raffelt, *Euro.Phys.J.C*36, 405-410 (2004).
7. N.Seiberg and E.Witten, *JHEP*032, 9909 (1999).
8. B. Melic,K.P.Kumericki,J.Trampetic,P.Schupp and M.Wohlgenannt,*Eur.Phys.J.C*42, 499-504 (2005)
9. OPAL Collab:*Phys.LettB*56,181-190,(2003); I.Hinchliffe,N.Kersting and Y.L.Ma,*IJMPA*.19, 179-204 (2004)
10. S. Baek, D. K. Ghosh, X.G. He and W-Y. P. Hwang, *Phys.Rev. D*64, 056001 (2001).
11. C.E. Carlson , C.D. Carone and N. Zobin, *Phys. Rev. D* 66, 075001 (2002).
12. FNAL E704 Collab.: *Phys. Lett. B* 261, 201-206 (1991); *Phys.Rev. Lett.* 77, 2626-2629 (1996).
13. G.Altarelli and G.Parisi, *Nucl.Phys.B*.126, 298-331(1977).

Twistor Strings, Gauge Theory and Gravity

Mohab Abou-Zeid

Institut des Hautes Etudes Scientifiques
35 route de Chartres, Le Bois-Marie, 91440 Bures-sur-Yvette, France

Abstract. The development of new analytic and computational methods in the perturbative approach to Yang-Mills theories is of great importance to the understanding of LHC physics. Twistor string theory, a form of string theory defined on Penrose's twistor space, may provide a very efficient and powerful alternative to the usual Feynman diagram expansion of field theory. I survey the recent construction of a family of new twistor string theories which are free from world-sheet anomalies and give the space-time spectra of Einstein supergravities, with second order field equations instead of the higher derivative conformal supergravities that arose from earlier twistor strings.

Keywords: Classical Theories of Gravity, Topological Strings, Superstrings and Heterotic Strings
PACS: 04.20.-q, 11.25.-w

INTRODUCTION

Within a couple of years from now, the Large Hadron Collider (LHC) will begin operation at CERN, providing proton-proton collisions at 14 TeV center of mass (COM) energy and opening a new window into physics at the shortest distance scales. The luminosity should be $\sim 10-100$ times greater than at Fermilab's Tevatron and the combined rise in energy and luminosity will hugely increase the production of particles with masses in the range 10^2 to 10^3 GeV, including electroweak vector bosons, top quarks, Higgs bosons, and perhaps new particles representing physics beyond the Standard Model (SM). In this context, a detailed quantitative understanding of both the new physics signals and the SM backgrounds is required.

The SM backgrounds for e. g. processes that may contain several jets and (perhaps) a few electroweak bosons are very complex, but a simplifying feature is that the masses of the observed final-state particles in these reactions are generally negligibly small (except for the cases of the W, Z, or top quark, which however immediately decay to essentially massless quarks or leptons). So if we include the decay processes in the description of the event, every final-state particle is approximately massless. We can also (usually) neglect the masses of the colliding partons (the quarks and gluons). Thus the backgrounds (and many signals) require a detailed understanding of scattering amplitudes for many ultrarelativistic (massless) particles, in particular of the quarks and gluons of quantum chromodynamics (QCD).

In principle, Feynman rules are all we need to evaluate the tree and loop amplitudes. In practice, although Feynman rules are very general and apply to any local quantum field theory, the standard method of drawing up all diagrams, computing them using the Feynman rules and then summing all the terms becomes extremely inefficient and cumbersome as the number of external legs grows. The development of new analytic and computational methods in the perturbative approach to Yang-Mills (YM) theories is thus of great and imminent experimental importance: without clear theoretical predictions from perturbative QCD, there would be difficulties in interpreting LHC data. Some new and exciting progress has recently been made in this direction using a variety of field theoretic and string theory-inspired techniques which provide very efficient and powerful alternatives to the usual Feynman diagram expansion. These gauge theory breakthroughs and many others were directly inspired by the recent development of *twistor string theory*, which is a new form of string theory defined on supertwistor space.

The main reasons for the inefficiency of the Feynman diagram expansion are:

- Too many diagrams: many diagrams are related by gauge invariance.

CP1006, *High Energy Physics and Applications*, edited by J. Ellis, S. Nasri, and E. Malkawi

- Too many terms in each diagram: nonabelian gauge boson self-interactions are complicated.
- Too many kinematic variables (allowing the construction of arbitrarily complicated expressions).

Consequently, intermediate expressions tend to be vastly more complicated than the final results, when the latter are represented in an appropriate way. In order to fix ideas, let me give two examples of this:

1. In QCD, the tree-level amplitudes for the scattering of n gluons all vanish if the helicities of the gluons (considered as outgoing particles) are either a) all the same b) all the same, except for one of opposite helicity. Using parity, we can take the bulk of the gluons to have positive helicity, and write this vanishing relation as

$$A_n^{tree}(1^\pm,2^+,3^+,\ldots,n^+)=0. \tag{1}$$

This vanishing is somewhat mysterious from the point of view of Feynman diagrams.

2. The first sequence of nonvanishing tree amplitudes has two gluons with negative helicity, labelled by j and l, say, and the rest of positive helicity. This sequence of maximally helicity-violating (MHV) amplitudes has the exceedingly simple Parke-Taylor (PT) form [1]:

$$\begin{aligned} A_n^{MHV,jl} &= A_n^{tree}(1^+,2^+,\ldots,j^-,\ldots,l^-,\ldots,n^+) \\ &= i\frac{\langle jl\rangle^4}{\langle 12\rangle\langle 23\rangle\ldots\langle n1\rangle} \end{aligned} \tag{2}$$

in terms of $SU(2)$ invariant spinor products

$$\langle ij\rangle = \varepsilon^{AB}\lambda_A\tilde{\lambda}_{A'} \qquad A=0,1 \qquad A'=0',1'. \tag{3}$$

Here we have conveniently written the massless momenta $k_i^2=0$ of the gluons as the product of a left-handed spinor with a right-handed one using the 2-spinor notation:

$$k_i^\mu(\sigma_\mu)_{AA'}=(\lambda_i)_A(\tilde{\lambda}_i)_{A'}. \tag{4}$$

The PT expression above gives the piece of the full amplitude in which the n gluons have a definite cyclic ordering; the full amplitude can be built out of permutations of such partial amplitudes. The formula is extremely simple, and in particular it is 'holomorphic' in the spinor variables, a fact which is very mysterious from the point of view of Feynman diagrams.

TWISTOR SPACE AND TWISTOR STRINGS

A breakthrough in the interpretation of such properties of YM scattering amplitudes came when Witten transformed the PT amplitude from the traditional momentum-space variables into Penrose's twistor space [2]. The twistor transform is a kind of Fourier transform, and there are many examples where transforming a problem into the right variables can expose its simplicity. In this case, Witten found a remarkable interpretation of the formula and of its 'holomorphy' properties in terms of a particular string theory on a particular supersymmetric (SUSY) version of twistor space.

Witten defined twistor string theory as a topological B-model with target the supermanifold $\mathbb{CP}^{3|4}$, aka $N=4$ projective supertwistor space. What is projective $N=4$ supertwistor space $\mathbb{CP}^{3|4}$? First, consider complexified flat spacetime $\mathbb{C}^4$. Spacetimes of signature (4,0), (3,1) or

(2,2) can be obtained by restricting to suitable real slices of the complexification. Complexified twistor space $\mathbb{T}$ is another copy of $\mathbb{C}^4$ with coordinates $Z^\alpha = (Z^0, Z^1, Z^2, Z^3)$. Projective twistor space $\mathbb{PT}$ is $\mathbb{CP}^3$ ($Z^\alpha \sim \lambda Z^\alpha$ for some $\lambda \in \mathbb{C}^*$) and a point Z in $\mathbb{PT}$ has homogeneous coordinates $[Z^0, Z^1, Z^2, Z^3]$. $\mathbb{T}$ can be defined as the representation space of spinors of the complexified conformal group $SL(4,\mathbb{C})$. The Z^α transform as a **4** of $SL(4,\mathbb{C})$ and decompose into 2-component spinors under the complexified Lorentz group $SL(2,\mathbb{C}) \times SL(2,\mathbb{C}) \subset SL(4,\mathbb{C})$; we write $Z^\alpha = (\omega^A, \pi_{A'})$ with $A = 0,1$ and $A' = 0', 1'$.

Twistor theory has many important and useful applications, including the Ward construction [3] solving the self-dual (SD) YM equations and Penrose's 'nonlinear graviton' construction [4] solving the self-dual Einstein equations. Most stem from the following basic relationship. A point $x^{AA'} \in \mathbb{CM}$ corresponds to a 2-dimensional linear subspace of $\mathbb{T}$ given by the incidence relation

$$\omega^A = x^{AA'} \pi_{A'} \tag{5}$$

and to a projective line (a copy of $\mathbb{CP}^1$) in $\mathbb{PT}$.

The basic correspondence between twistor space and spacetime is:

$\mathbb{PT}$	$\mathbb{CM}$
complex projective line	point
point	alpha-plane
intersection of lines	null separation of points

Complexified Minkowski space $\mathbb{CM}$ can be thought of as the moduli space of complex projective lines, while (projective) twistor space $\mathbb{PT}$ can be thought of as the moduli space of α-planes [5].

One remarkable application of twistor space is the Penrose transform [5]. This identifies solutions to (massless) free field equations for spacetime fields of arbitrary helicity $-\frac{n}{2}$ on a suitable region $U \subset \mathbb{CM}$ with the cohomology group $H^1(\mathbb{PT}(U), \mathcal{O}(n-2))$, where $\mathbb{PT}(U) \subset \mathbb{PT}$ is the corresponding subset of $\mathbb{PT}$. The $N = 4$ SUSY generalisation of the above constructions is more or less straightforward [2, 9].

The B-model on $\mathbb{CP}^{3|4}$ (or any Calabi-Yau target) describes – for open strings – holomorphic bundles and more general sheaves, together with their moduli. Via the Penrose transform, the open strings (for 'space-filling branes' [2]) reproduce the perturbative spectrum of $N = 4$ super YM (SYM). The interactions can also be reproduced, although this is somewhat harder to see. For closed strings the vertex operator for the physical $(0,1)$-form on $\mathbb{CP}^{3|4}$ describes linearised deformations of the complex structure of a suitable region of $\mathbb{CP}^{3|4}$ (e. g. a neighbourhood $\mathbb{PT}_0$ of a projective line in $\mathbb{CP}^{3|4}$). Penrose showed that such complex structure deformations lead to nontrivial deformations of the flat twistor correspondence wherein space-time is deformed from flat space to one with a curved conformal structure with anti-selfdual Weyl curvature. The construction provides a correspondence between curved twistor spaces and conformally anti-selfdual space-times, yielding a general construction of such space-times. This 'nonlinear graviton' construction establishes that space-time $\mathbb{C}\mathscr{M}$ together with its anti-self-dual conformal structure can be reconstructed from the complex structure of curved twistor space $\mathscr{T}$ together with (Υ, Ω), where Υ denotes the Euler homogeneity operator and Ω the holomorphic 3-form, or from $P\mathscr{T}$ and its complex structure. The existence of the correspondence is preserved under small deformations, either of the complex structure on $P\mathscr{T}$, or of the anti-self dual conformal structure on $\mathbb{C}\mathscr{M}$.

In Witten's twistor string theory, the twistor space string field theory action has a term with a Lagrange multiplier imposing $N(J) = 0$, where $N(J)$ is the Nijenhuis tensor of the complex structure J of the deformed region. Via the nonlinear graviton construction, these integrable complex structure deformations of e. g. $\mathbb{PT}_0$ describe solutions of the ASD Weyl equations in spacetime:

$$W_{ABCD} = 0 \tag{6}$$

where $W_{\mu\nu\rho\sigma}$ denotes the Weyl tensor with SD and anti-SD (ASD) parts W_{ABCD} and $W_{A'B'C'D'}$. This describes one helicity of conformal gravity, and leads to an action of the form

$$\int d^4x \sqrt{g} U^{ABCD} W_{ABCD} \tag{7}$$

with U symmetric in all its indices and of Lorentz spin $(2,0)$. In addition there is a term

$$\int d^4x \sqrt{g} U^{ABCD} U_{ABCD} \tag{8}$$

which arises from D-instantons in Witten's topological B-model. Integrating out U gives an action which is equivalent to the conformal gravity action

$$\int d^4x \sqrt{g} W^{\mu\nu\rho\sigma} W_{\mu\nu\rho\sigma} \tag{9}$$

(the difference between the two being a topological term). Thus one gets both the spectrum and the interactions of $N=4$ conformal supergravity (CSUGRA) [6].

Similar results can be derived using an alternative formulation of the twistor string due to Berkovits [7, 8]. The Berkovits model reproduces the correct tree-level SYM amplitudes using ordinary string tree amplitudes as opposed to D-instanton contributions. The construction uses the fact that in split spacetime signature $++--$ there is a 3-real dimensional submanifold $P\mathscr{T}_{\mathbb{R}}$ of complex twistor space, e. g. the standard embedding $\mathbb{RP}^3 \subset \mathbb{CP}^3$ in the flat case, and the data about deformations of the complex structure is encoded in an analytic vector field f on $P\mathscr{T}_{\mathbb{R}}$.

The Berkovits model is a theory of open strings with boundaries on $\mathbb{PT}_{\mathbb{R}}$ and action

$$S = \int_{\Sigma} d^2\sigma \left(Y_I \bar{\partial} Z^I + \tilde{Y}_J \partial \tilde{Z}^J - \tilde{A} J - A\tilde{J} \right) + S_C. \tag{10}$$

Here the Y's are variables conjugate to the $Z^I = (Z^\alpha, \psi^a)$ (with the ψ^a, $a = 1, \ldots 4$ fermionic coordinates on the 'soul' of $\mathbb{CP}^{3|4}$), A is a $GL(1,\mathbb{C})$ world-sheet gauge field and J is the associated conserved current on the world-sheet Σ. The $GL(1,\mathbb{C})$ invariance insures that the theory is defined on projective twistor space $\mathbb{PT}$ rather than on twistor space $\mathbb{T}$. The world-sheet theory includes a left and right-moving current algebra S_C reminiscent of heterotic string constructions, and this plays a key role both in the quantum consistency of the model and in determining its spacetime symmetries. In particular, it insures cancellation of conformal anomalies: in addition to the usual Virasoro ghost (b,c) system with $c = -26$ there are ghosts (u,v) and $(\tilde{u},\tilde{v})$ for the $GL(1,\mathbb{C})$ symmetry of the action which contribute $c = -2$.

$N=4$ SYM physical states are created by the dimension one $GL(1,\mathbb{C})$-neutral primary field

$$V_\phi = j_r \phi^r(Z) \tag{11}$$

where $\phi^r(Z)$ are functions of $GL(1)$ charge zero and the j_r with $r = 1, \ldots \dim G$ are the (left-moving) Kač-Moody algebra currents. G becomes a spacetime SYM gauge group. The superfields $\phi(Z)$ are functions of homogeneity degree zero and have an expansion of the form $\phi(Z) = A_1 + \ldots + A_{-1}\psi^4$, which gives an $N=4$ SYM multiplet via the Penrose transform.

Similarly (I will not give the details here) $N=4$ CSUGRA physical states are created by an open string vertex operator constructed from a vector field f defined on $\mathbb{PT}_{\mathbb{R}}$, corresponding to deformations of the embedding of $\mathbb{PT}_{\mathbb{R}}$ in $\mathbb{PT}$, together with a vertex operator constructed from a 1-form $g_I dZ^I$ on $\mathscr{T}$:

$$V_f = Y_I f^I(Z) \qquad\qquad V_g = g_I(Z) \partial Z^I. \tag{12}$$

The physical state conditions are

$$\partial_I f^I = 0 \qquad\qquad Z^I g_I = 0 \tag{13}$$

and the gauge invariances are

$$\delta f^I = Z^I \Lambda \qquad\qquad \delta g_I = \partial_I \chi. \tag{14}$$

The emergence of CSUGRA is disappointing from the point of view of Witten's original goal of describing pure $N=4$ SYM: e. g. at tree level the topological B-model computes the amplitudes of $N=4$ SYM as desired [2], but at loops it computes the amplitudes of $N=4$ SYM conformally coupled to $N=4$ CSUGRA [6]. Moreover there does not seem to be any obvious limits which one could take in the B-model or in the Berkovits model in order to decouple the gauge and gravitational sectors. So CSUGRA seems unavoidable in the Witten and Berkovits models, but this theory is generally considered to be inconsistent: it leads to 4th order PDE for the fluctuations of the metric, and thus to a lack of unitarity.

On the other hand, $N=4$ SYM makes sense without CSUGRA, and it would be desirable to find a perturbative string theory description of it. A twistor string that gave Einstein SUGRA (with 2nd order field equations for the graviton) coupled to SYM would be much more useful, and might have a limit in which the gravity could be decoupled. The spacetime conformal invariance would be broken in such a theory, and in particular this would introduce a dimensionful parameter which could be used to define the decoupling limit.

NONLINEAR GRAVITON AND NEW TWISTOR STRINGS

In fact there is an important variant of the Penrose construction that applies to conformally SD spaces that are also Ricci-flat, so that the full Riemann tensor is self-dual:

$$W_{A'B'C'D'} = 0, \qquad R_{\mu\nu} = 0 \qquad \Rightarrow \qquad R_{A'B'C'D'} = 0. \tag{15}$$

The corresponding twistor spaces $P\mathscr{T}$ then have extra structure. In particular, they have a fibration $P\mathscr{T} \rightarrow \mathbb{CP}^1$. The holomorphic 1-form on $\mathbb{CP}^1$ pulls back to give a holomorphic 1-form on $P\mathscr{T}$ which takes the form $I_{\alpha\beta} Z^\alpha dZ^\beta$ in homogeneous coordinates Z^α, for some $I_{\alpha\beta}(Z) = -I_{\beta\alpha}(Z)$ (which are the components of a closed 2-form on the non-projective twistor space $\mathscr{T}$). The dual bi-vector $I^{\alpha\beta} = \frac{1}{2}\varepsilon^{\alpha\beta\gamma\delta} I_{\gamma\delta}$ defines a Poisson structure and is called the *infinity twistor*. Choosing a point at infinity, corresponding to such an infinity twistor, breaks the conformal group down to the Poincaré group; e. g. on Minkowski space, the infinity twistor determines the light-cone at infinity in the conformal compactification. A similar situation obtains more generally: the infinity twistor breaks conformal invariance.

Self-dual space-times are obtained by seeking deformations of the complex structure of twistor space as before, but now Ricci-flatness in space-time places further restrictions on the deformations allowed [4, 10]. The vector field f on $\mathbb{RP}^3$ is required to be a Hamiltonian vector field with respect to the infinity twistor:

$$f^\alpha = I^{\alpha\beta} \frac{\partial h}{\partial Z^\beta} \tag{16}$$

for some function h of homogeneity degree 2 on $\mathbb{RP}^3$. In the linearised theory, such a function h corresponds to a positive-helicity graviton in space-time via the Penrose transform, and the non-linear graviton construction gives the generalisation of this to the non-linear theory [4, 10].

The existence of the nonlinear graviton construction suggests seeking twistor strings that are modifications of the Berkovits or the Witten model with explicit dependence on the infinity twistor, such that there are extra constraints on the vertex operators imposing that the deformation of the complex structure be of the form given above. Then the leading term in the action analogous to

$$\int d^4x \sqrt{g} U^{ABCD} W_{ABCD} \tag{17}$$

should have a multiplier imposing self-duality, not just conformal self-duality, and further terms quadratic in the multiplier (from instantons in Witten's approach) could then give Einstein gravity.

A formulation of Einstein gravity of just this form was discussed in [12], and an $N = 8$ SUSY version was constructed some time ago by Siegel [13]. In [9], we constructed new twistor string models which appear to give Einstein (super)gravity coupled to (super) Yang-Mills. The new theories are constructed by gauging certain symmetries of the Berkovits twistor string which are defined by the infinity twistor $I^{\alpha\beta}$ (or its appropriate SUSY generalisation I^{IJ}). Their structure is very similar to that of the Berkovits model, but the gauging adds new terms to the BRST operator so that the vertex operators have new constraints and gauge invariances. I will shortly describe the spectra and some of the interactions of 2 classes of theories for which the world-sheet anomalies cancel.

The corresponding target space theories can be expected to be anomalous in general, with the anomalies arising from inconsistencies in the corresponding twistor string model, though the mechanism for this is as yet unknown. This may rule out some of the models we construct, or restrict the choice of gauge group G. The situation is similar to that of the Berkovits and Witten models, which give target space theories that are anomalous in general, with the anomalies canceling only for the 4-dimensional groups $G = U(1)^4$ or $G = SU(2) \times U(1)$.

The 1st class of anomaly-free theories is formulated in $N = 4$ supertwistor space. Gauging a symmetry of the string theory generated by 1 bosonic and 4 fermionic currents gives a theory with the spectrum of $N = 4$ Einstein SUGRA coupled to $N = 4$ SYM with arbitrary gauge group G. Gauging a single bosonic current gives a theory with the spectrum of $N = 8$ Einstein SUGRA, provided the number of $N = 4$ vector multiplets is 6. In the YM sector, the string theory is identical to that of Berkovits, so that it gives the same tree level YM amplitudes. Both gauged theories have the MHV 3-graviton interaction (with 2 positive helicity gravitons and 1 negative helicity one) of Einstein gravity. The other interactions are still being computed, and the results should determine the form of the interacting theories [14].

There are different interacting theories with the spectrum of $N = 4$ Einstein SUGRA (coupled to $N = 4$ SYM): the standard non-chiral Einstein SUGRA and Einstein SUGRA with chiral interactions. For the theory with $\dim G = 6$ and the spectrum of $N = 8$ SUGRA, the interactions could be those of the standard $N = 8$ SUGRA or those of Siegel's chiral $N = 8$ SUGRA.

The 2nd class of string theories is obtained by gauging different numbers of bosonic and fermionic symmetries so that anomalies are cancelled against ghost contributions for strings in twistor spaces with 3 complex bosonic dimensions and any number N of complex fermionic dimensions, corresponding to theories in 4-dimensional spacetime with N supersymmetries. Analysing the spectrum of states arising from ghost-independent vertex operators, one finds:

- For $N = 0$, a theory with the bosonic spectrum of SD gravity together with SD YM and a scalar.
- For $N < 4$, supersymmetric versions of this $N = 0$ SD theory.
- For $N = 4$, a (second) theory with the spectrum of $N = 4$ Einstein SUGRA coupled to $N = 4$ SYM with arbitrary G.

Consistent non-linear interactions are possible classically for the $N = 0$ theory. The field equations are given by a scalar-dependent modification of the equations for SD gravity coupled to SD YM, and the (noncovariant) action is presumably of the Plebanski type. This $N = 0$ theory may be closely related to the interacting theory of SD gravity coupled to SD YM arising from the (Ooguri-Vafa) $N = 2$ world-sheet supersymmetric string [11]. The theories with $N \leq 4$ are supersymmetric extensions of the $N = 0$ theory, and could be consistent at the interacting level if the $N = 0$ theory is. The determination of the precise form of the interactions must await a detailed investigation of the scattering amplitudes in the new twistor string models [14].

An important point here is that if one scales the infinity twistor $I^{IJ} \to RI^{IJ}$, $\varepsilon^{AB} \to R\varepsilon^{AB}$, while keeping f^I, g_I fixed, then $h \to R^{-1}h$ and $\tilde{h} \to R\tilde{h}$. Then the amplitude scales as R^{-1}, so that R^{-1} sets the strength of the gravitational coupling. Thus there is a decoupling limit of gravity in our models, which should be useful in computing YM amplitudes at loops using the twistor string. In the case of the model giving the spectrum (and some of the interactions) of $N = 8$ SUGRA, we hope to use the string theory to calculate $N = 8$ SUGRA loop amplitudes as well as to investigate the conjectured ultra-violet finiteness of this very special SUSY field theory of gravity, and the fascinating and intimate connection of its scattering amplitudes to those of $N = 4$ SYM.

ACKNOWLEDGEMENTS

The author thanks Christopher Hull and Lionel Mason for very enjoyable collaborations, and the organizers of the UAE-CERN workshop in Al Ain for their hospitality and for financial support.

REFERENCES

1. S. Parke and T. Taylor, *An amplitude for N gluon scattering*, Phys. Rev. Lett. **56** (1986) 2459.
2. E. Witten, *Perturbative gauge theory as a string theory in twistor space*, Commun. Math. Phys. **252** (2004), [arXiv:hep-th/0312171].
3. R. S. Ward, *On selfdual gauge fields*, Phys. Lett. A **61** (1977) 81.
4. R. Penrose, *Nonlinear gravitons and curved twistor theory*, Gen. Rel. Grav. **7** (1976) 31.
5. S. Huggett and K. Tod, *An introduction to twistor theory*, Second Edition, Cambridge University Press, Cambridge (1994).
6. N. Berkovits and E. Witten, *Conformal supergravity in twistor-string theory*, JHEP **0408** 009 (2004), [arXiv:hep-th/0406051].
7. N. Berkovits, *An alternative string theory in twistor space for N = 4 super-Yang-Mills*, Phys. Rev. Lett. **B93** 011601 (2004), [arXiv:hep-th/0402045].
8. N. Berkovits and L. Motl, *Cubic twistorial string field theory*, JHEP **0404** (2004) 056, [arXiv:hep-th/0403187].
9. M. Abou-Zeid, C. M. Hull and L. J. Mason, *Einstein supergravity and new twistor string theories*, [arXiv:hep-th/0606272].
10. L. J. Mason, *The relationship between spin-2 fields, linearized gravity and linearized conformal gravity*, Twistor Newsletter **23** (1987) 67 reprinted as § I.2.18 *Further advances in twistor theory, Volume I: The Penrose transform and its applications*, L. J. Mason and L. P. Hughston eds., Pitman research notes in Maths **231**, Longman (1990).
11. H. Ooguri and C. Vafa, *Geometry of N=2 strings*, Nucl. Phys. B **361**, 469 (1991).
12. M. Abou-Zeid and C. M. Hull, *A chiral perturbation expansion for gravity*, JHEP **0602**, 057 (2006), [arXiv:hep-th/0511189].
13. W. Siegel, *Selfdual N=8 supergravity as closed N=2 (N=4) strings*, Phys. Rev. D **47** (1993) 2504, [arXiv:hep-th/9207043].
14. M. Abou-Zeid, C. M. Hull and L. J. Mason, in preparation.

Quaternionic Analysis of Generalized Electromagnetic Fields of Superluminal Dyons

P. S. Bisht and O. P. S. Negi

March 18, 2008

Department of Physics,Kumaun University,S. S. J. Campus,Almora - 263601 (India)
Email:- ps_bisht123@rediffmail.com; ops_negi@yahoo.co.in

Abstract

Superluminal electromagnetic fields of dyons are described in T^4- space and Quaternion formulation of various quantum equations is derived. It is shown that on passing from subluminal to superluminal realm via quaternion the theory of dyons becomes the Tachyonic dyons. Corresponding field Equations of Tachyonic dyons are derived in consistent, compact and simpler form.
Keywords: Quaternion, dyons, Tachyons, Superluminal Transformations and electromagnetic fields.
PACS: 14.80 Hv.

In order to understand the existence of magnetic monopoles [1] and dyons [2], a self consistent quantum field theory of generalized electromagnetic fields associated with dyons (particle carrying electric and magnetic charges) of various spins has already been developed [3, 4] in terms of two four potentials [5] where the generalized charge, generalized four-potential, generalized field tensor, generalized vector field and generalized four-current density associated with dyons are described as complex quantities with their real and imaginary parts as electric and magnetic constituents. On the other hand, there has been continuing interest [6] in higher dimensional kinematical models for proper and unified theory of subluminal (bradyonic) and superluminal (tachyonic) objects [7]. The problem of representation and localizations of superluminal objects (particles) has been solved [8] in terms of higher dimensional space-time and it has been shown that the localization space for superluminal objects (tachyons) is T^4- space with one space and three times while that for subluminal objects (bradyons) is R^4- space (three space and one time). The concepts of superluminal Lorentz transformations, need of higher dimensional space-time, localizability of bradyons and tachyons are accordingly introduced [9] to under take the study of generalized fields of dyons under superluminal Lorentz transformations (*SLTs*). As such, the

CP1006, *High Energy Physics and Applications*, edited by J. Ellis, S. Nasri, and E. Malkawi

resulting space for bradyons and tachyons has been identified as the R^6- or $M(3,3)$ space where both space and time and energy and momentum are considered as vector quantities. So, we have two four dimensional slices of R^6- or $M(3,3)$ space; a subluminal (bradyonic) [8, 9]space $M(1,3) = [t,x,y,z](R^4 - space)$ where the coordinates t_x and t_y are not observed or coupled together giving $t = (t_x^2 + t_y^2 + t_z^2)^{\frac{1}{2}}$ and a superluminal (tachyonic) [8] space $M'(1,3) = [t_{z'}, x', y', z'] = [r, t_x, t_y, t_z](T^4 - space)$ where x and y are not observed or coupled together giving rise to $r = (x^2 + y^2 + z^2)^{\frac{1}{2}}$. Hence, a bradyonic $R^4 = M(1,3)$ space now maps to a tachyonic $T^4 = M'(3,1)$ space or vice versa [8]i.e. $R^4 = M(1,3) \overset{SLT}{\rightarrow} M'(3,1) = T^4$ where SLT denotes superluminal transformations. In T^4- space, the space-momentum are scalars while time-energy are vector quantities when viewed from subluminal frame of reference. Similar mapping also exists for four-potential, four currents and other four vectors. As such, the components of generalized four electromagnetic potential of dyons in R^4- space maps to a tachyonic $T^4 = M'(3,1)$ space as $(\overrightarrow{V}, i\phi) \overset{SLT}{\rightarrow} (\overrightarrow{\phi}, iV)$ where $\phi = |\overrightarrow{\phi}| = \phi_x^2 + \phi_y^2 + \phi_z^2)^{\frac{1}{2}}$ and $V = |\overrightarrow{V}| = (V_x^2 + V_y^2 + V_z^2)^{\frac{1}{2}}$on passing from subluminal to superluminal realm via SLTS.

The generalized four potential $\{\phi_\mu\} = \{\overrightarrow{\phi}, iV\}$ associated with superluminal (tachyonic) thus direcly leads to the following expressions for generalized electromagnetic fields of dyons in T^4-space i.e.

$$\overrightarrow{E_T} = -\overrightarrow{\nabla}_t A - \frac{\partial \overrightarrow{\phi_e}}{\partial r} - \overrightarrow{\nabla}_t \times \overrightarrow{\phi_m} \quad ; \quad \overrightarrow{H_T} = -\overrightarrow{\nabla}_t B - \frac{\partial \overrightarrow{\phi_m}}{\partial r} + \overrightarrow{\nabla}_t \times \overrightarrow{\phi_e} \tag{1}$$

where $\overrightarrow{\phi} = \overrightarrow{\phi_e} - i\overrightarrow{\phi_m}; V = A - iB (i = \sqrt{-1}$ are the components of vector and scalar parts of generalized four potential whose real and imaginary parts are respectively described as the electric and magnetic constituents of superluminal dyons. The vector wave function $\overrightarrow{\psi_T} == \overrightarrow{E_T} - i\overrightarrow{H_T}$ associated with generalized electromagnetic fields leads the following pair of Generalized Maxwell's Dirac (GDM) equations for generalized fields of dyons in T^4- space (for tachyonic dyons via superluminal transformation) i.e.

$$\overrightarrow{\nabla}_t . \overrightarrow{\psi_T} = \Im_0 \quad ; \quad \overrightarrow{\nabla}_t \times \overrightarrow{\psi_T} = -i\overrightarrow{\Im} - i\frac{\partial \overrightarrow{\psi_T}}{\partial r} \tag{2}$$

where $\Im_0$ and $\overrightarrow{\Im}$ are the components of generalized four - current source densities of dyons in T^4 - space , given by $\{\rho_\mu\} = \{\rho_\mu^e\} - i\{\rho_\mu^m\} = \{\Im_0, \overrightarrow{\Im}\}$. On using relation (1), we get the following expression for generalized vector field in terms of the components of generalized four potential of dyon in T^4 - space i.e.

$$\overrightarrow{\psi_T} = -\frac{\partial \overrightarrow{\phi}}{\partial r} - \overrightarrow{\nabla}_t V - i\overrightarrow{\nabla}_t \times \overrightarrow{\phi}. \tag{3}$$

Accordingly, we write the following tensorial forms of generalized Maxwell's - Dirac equations of dyons under the influence of superluminal transformation (in T^4- space) i.e.

$$f_{\mu\nu,\nu} = \rho_\mu^e; \ \widetilde{f_{\mu\nu,\nu}} = \rho_\mu^m \tag{4}$$

where $f_{\mu\nu}$ and $\widetilde{f_{\mu\nu}}$ are described as superluminal electric and magnetic fields tensors in T^4 - space giving rise to Generalized Maxwell's Dirac (GDM) equations coupled to electric and magnetic four currents in T^4- space. Similarly we may obtain a new vector parameter (current) $\overrightarrow{S_T}$, in T^4- space as

$$\overrightarrow{S_T} = \Box_t \overrightarrow{\psi_T} = \frac{\partial \overrightarrow{\mathfrak{I}}}{\partial r} - \overrightarrow{\nabla_t}\mathfrak{I}_0 - i\overrightarrow{\nabla_t} \times \overrightarrow{\mathfrak{I}} \quad (5)$$

where $\Box_t = \partial_r^2 - \nabla_t^2 = \frac{\partial}{\partial r^2} - \frac{\partial}{\partial t_x^2} - \frac{\partial}{\partial t_y^2} - \frac{\partial}{\partial t_z^2}$. We may now write the above said quantum equations of dyons in terms of simple and compact quaternion notations.

A quaternion is defined as $q = q_0 e_0 + q_1 e_1 + q_2 e_2 + q_3 e_3$ where $q_\alpha(\alpha = 0,1,2,3)$ are real or complex numbers with three imaginary units e_1, e_2 and e_3 satisfy the relations $e_i e_j = -\delta_{ij} + \varepsilon_{ijk} e_k$. The set of quaternions is called a 4 - dimensional real space essentially because 4- real numbers are required to specify a quaternion. The complex conjugate of q i.e. $q^\star$ is defined as $q^\star = q_0 + q_1^\star e_1 + q_2^\star e_2 + q_3^\star e_3$ while the quaternion conjugate $\overline{q} = q_0 - q_1 e_1 - q_2 e_2 - q_3 e_3$. We have $(qp)^\star = q^\star p^\star$ and $(\overline{qp}) = \overline{p}\,\overline{q}$ where p and q , are also the quaternions showing that the quaternion conjugate of product of two quaternions is the product of conjugates in reverse order. The scalar product of a two quaternions is defined as $(p,q) = \frac{1}{2}(p\overline{q} + q\overline{p})$; the norm of a quaternion is $|q| = (q.q) = q_0^2 + q_1^2 + q_2^2 + q_3^2$ and the inverse of a quaternion q is also a quaternion given by $q^{-1} = \frac{q}{|q|^2}$. As such, it is easy to write a four vector in T^4- space as a quaternion.

Let us write the generalized four-potential and generalized four - current associated with superluminal dyons in T^4 -space as quaternions i.e.

$$\phi = -iV + e_1\phi_1 + e_2\phi_2 + e_3\phi_3 \quad (6)$$

$$\rho = -iJ + e_1\rho_1 + e_2\rho_2 + e_3\rho_3. \quad (7)$$

In this case the quaternionic differential operator is written as

$$\boxdot_t = -i\partial_r + e_1\partial_1 + e_2\partial_2 + e_3\partial_3 \quad (8)$$

where $\partial_r = \frac{\partial}{\partial r}, \partial_1 = \frac{\partial}{\partial t_x}, \partial_2 = \frac{\partial}{\partial t_y}, \partial_3 = \frac{\partial}{\partial t_z}$. Operating equation (8) on equations (6) and (7) and using the quaternion multiplication rules , we get

$$\boxdot_t \phi = \psi_T \quad ; \quad \boxdot_t \rho = S_T \quad (9)$$

where $\psi_T = -\psi_r - i(e_1\psi_1 + e_2\psi_2 + e_3\psi_3$ and $S_T = -S_r - i(e_1 S_1 + e_2 S_2 + e_3 S_3$ with $\psi_r = \partial_r V + \partial_1\phi_1 + \partial_2\phi_2 + \partial_3\phi_3 = 0$ and $S_r = \partial_r J + \partial_1\rho_1 + \partial_2\rho_2 + \partial_3\rho_3 = 0$. Equations (9) are the quaternionic differential equations for generalized potential and generalized current of superluminal dyons in T^4- space. The conjugate quaternion field equations in T^4 - space may then be

expressed as,

$$\overline{\boxdot_t}\,\overline{\phi} = \overline{\psi_T} \qquad ; \qquad \overline{\boxdot_t}\,\overline{\rho} = \overline{S_T} \tag{10}$$

where $\overline{\boxdot_t} = -i\partial_r - (e_1\partial_1 + e_2\partial_2 + e_3\partial_3); \overline{\phi} = -iV - (e_1\phi_1 + e_2\phi_2 + e_3\phi_3)$ **;** $\overline{\rho} = -iJ - (e_1\rho_1 + e_2\rho_2 + e_3\rho_3)$**;** $\overline{\psi_T} = -\psi_r + i(e_1\psi_1 + e_2\psi_2 + e_3\psi_3)$ **and** $\overline{S_T} = -S_r - i(e_1S_1 + e_2S_2 + e_3S_3)$ **are described as the corresponding quaternion conjugates.**

As such the quaternionic formalism is thus compact, simpler, unique and consistent. It is also manifestly covariant under quaternion Lorentz transformations. It may be concluded that while passing from usual four space to quaternionic formulation the signature of four-vector is changed from $(+,+,+,-)$ to $(-,-,-,+)$ and the present quaternion analysis of generalized electromagnetic fields of superluminal dyons reproduces the theories of electric (magnetic) charge in the absence of magnetic (electric) charge or vice versa on dyons in T^4- space.

ACKNOWLEDGMENT

This work is carried out under the U-COST research scheme No. UCS&T/R&D/ PHY-03/06-07/1233.

References

[1] P. A. M. Dirac, *Proc. Roy. Soc. London* **A133,** 60 (1931).

[2] J. Schwinger, *Science* **165,** 757 (1978).

[3] B. S. Rajput and D. C. Joshi, *Had. J.* **4,** 1805 (1981).

[4] Shalini Bisht, P. S. Bisht and O. P. S. Negi, *IL Nuovo Cimento* **B113**, 1449 (1998) .

[5] N. Cabibo and E. Ferrari, *IL Nuovo Cimento* **23,** 1147 (1962).

[6] E. Recami, *Found. Phys.* **17,** 239 (1987) .

[7] O. M. P. Bilnaiuk, V. K. Deshpande and E. C. G. Sudarshan, *Amer. J. Phys.* **30,** 718 (1962).

[8] M. C. Pant, P. S. Bisht, O. P .S. Negi and B. S. Rajput, *Cand. J. Phys.* **78,** 303 (2000) .

[9] P. S. Bisht, Jivan Singh and O. P. S. Negi, "*Quaternionic Reformulation of Generalized Superluminal Electromagnetic Fields of Dyons*", arXiv:hep-th/0607212.

Indirect search for the fourth genration of quarks and leptons

Vali Bashiry

Engineering Faculty, Cyprus International University, Via Mersin 10, Turkey

Abstract. It is known that the Standard Model (SM) does not predict the number of fermion families; N. The only restriction comes from the asymptotic freedom of QCD which requires the number of quarks to be less than 17 and, therefore, the number of SM families to be $N \leq 8$. Before 1990ÂŠs, many authors published articles related to the extra SM families and their phenomenological consequences. In early 1990ÂŠs, the LEP data yields N is almost 3 where the neutral lepton mass for each family is less than half the mass of the Z boson. Generally, this result is interpreted as the exact value of N, since one assumes that the neutrinos must have very small masses. If we disregard this incorrect assumption, the LEP data does not exclude the existence of extra SM families with heavy neutrinos. Meanwhile, few papers arguing the existence of the fourth SM family have been publishing. These arguments are based on the ÂŚflavor democracyÂŠ hypothesis. The study of FCNC in B decays can indirectly shed light on the SM4 family in quark sector and LFV is a good candidate to indirect search for 4th generation of leptons.

Keywords: Fourth Generation, Rare decay, B Physics
PACS: 12.60.–i, 13.30.–a, 14.20.Mr

INTRODUCTION

In spite of the incredible successes of standard model, many fundamental issues remain: the replication of families and how many quarks and leptons are there? QCD does not throw any light on how many quarks flavors there should be? Origin of the masses and mixing which appear as free parameters since CKM matrix and Higgs couplings with fermions contain as many arbitrary constants; are some unanswered questions. Although there is a mechanism to incorporate CP violation in standard model with three generation(SM3) of quarks and leptons, it is still insufficient to explain baryogenesis.

The existence of the 4th generation of fermions, if their mass is less that the half of the mass of the Z boson, is excluded by the LEP II experiment[1]. In this sense, the status of fourth generations is more subtle [2] from the experimental point of view. However, a consequential extension of the SM3 may address some of the puzzles and fundamental questions from the theoretical point of view. In this regard, the consequential 4th generation of quarks and leptons attracts many attentions i.e., [3]. Also, heavy neutrinos of the 4th generation is searched for dark matter constituent[4]. The 4th generation of quarks can include new weak phases and mixing in CKM matrix. Thus, it may demonstrate better solution to the baryogenesis than the SM3.

Two type of studies can be conducted to discover the 4th generation of fermions. First is the direct search of the 4th generation of quarks and leptons which can be access by increasing the center of mass energy of colliders with high luminosity. Here, the cross

CP1006, *High Energy Physics and Applications*, edited by J. Ellis, S. Nasri, and E. Malkawi

section of production will increase and such fermions can be created as a real states. i.e., the 4th generation of quarks can be created by gluon–gluon fusion at LHC[5]. Second is the indirect search, which studies the effects of the 4th generation fermion in the FCNC decays and LFV [3]. In these classes of studies one study the contribution of the 4th generation fermions at the quantum loop level i.e., Wu-Jun Hou and Tai-Fu Feng [3] studied the effects of the 4th generation heavy neutrino(heavier than the half of the Z boson mass) in the $\mu \to e\gamma$ decay and on anomalous magnetic moment of the μ. They obtained that the experimental measurement makes upper limit to the mass of ν_4 up to $\sim$ 100GeV. With this constrain, one may study the branching ratio of the $\mu^- \to 2e^-e^+$ decay.

$b \to s(d)$ transition is forefront for searching the 4th generation quarks. This transition is forbidden at tree level in the Standard model with three generations. A consequential extension of third-generation standard model to the fourth-generation standard model will maintain the same property at three level, but at the quantum loop level the 4th generation heavy quark(t') can contribute to the quantum loop. This contribution can affects physical observables. i.e., branching ratio, CP asymmetry, polarization asymmetries and FB asymmetries. The study of these physical observables are good tools to look for the 4th generation up type quarks[3].

In this study, we try to examine the possible effects of the 4th generation in the polarized CP asymmetries of the inclusive $b \to s\ell^+\ell^-$ and polarized forward–backward asymmetries in $B \to K^*\ell^+\ell^-$ decay.

The paper includes 3 sections: In section 2, we modify the effective Hamiltonian in the presence of the 4th generation and polarized CP and FB asymmetries are presented, respectively. Section 3, devoted to the numerical analysis and conclusion.

GENERAL FORMALISM

The QCD corrected effective Lagrangian for the decays $b \to s\ell^+\ell^-$ can be achieved by integrating out the heavy quarks and the heavy electroweak bosons reads as:

$$M = \frac{G_F \alpha \lambda_t}{\sqrt{2}\pi}\Big[\; C_9^{\rm tot}(\bar{s}\gamma_\mu P_L b)\bar{\ell}\gamma_\mu \ell + C_{10}^{\rm tot}(\bar{s}\gamma_\mu P_L b)\bar{\ell}\gamma_\mu \gamma^5 \ell$$
$$-2C_7^{\rm tot}\bar{s} i\sigma_{\mu\nu}\frac{q^\nu}{q^2}(m_b P_R + m_s P_L)b\bar{\ell}\gamma_\mu \ell\Big], \qquad (1)$$

In writing this, unitarity of the CKM matrix has been used and the Wilson coefficients is extended to include the 4th generation effects as follows:

$$\lambda_t C_i \to \lambda_t C_i^{SM} + \lambda_{t'} C_i^{\rm new}, \qquad \text{where } \lambda_f = V_{fb}^* V_{fs}. \qquad (2)$$

The unitarity of the 4×4 CKM matrix leads to $\lambda_u + \lambda_c + \lambda_t + \lambda_{t'} = 0.0$ Since $\lambda_u = V_{ub}^* V_{us}$ is negligible in strength compared to the others . Then, $\lambda_t \approx -\lambda_c - \lambda_{t'}$.

$\lambda_{t'}$ can be parameterized as: $\lambda_{t'} = V_{t'b}^* V_{t's} = r_{sb} e^{i\phi_{sb}}$. The explicit forms of the $C_i^{\rm new}$ can be obtained from the corresponding expression of the Wilson coefficients in the SM by substituting $m_t \to m_{t'}$. The analytic expressions for all Wilson coefficients, except C_9^{eff}, can be found in [6]. The matrix element for the exclusive $B \to K^* \ell^+ \ell^-$ decay can be obtained by sandwiching the Eq. (1) over meson states,which is parameterized in terms of form factors[7].

Using the form factors, the matrix element of the $B \to K^* \ell^+ \ell^-$ decay is given Aliev*et al.*[3]. The spin direction of a lepton can be described by setting a reference frame with three orthogonal unit vectors S_L, S_N and S_T, such as:

$$S_L = \frac{p^-}{|p^-|}, \qquad S_N = \frac{p_d \times p^-}{|p_d \times p^-|}, \qquad S_T = S_N \times S_L ,$$

where p_d and p^- are the three momentum vectors of the s quark and the ℓ^- lepton, respectively, in the $\ell^+ \ell^-$ center-of-mass(CM) system. For a given lepton ℓ^- spin direction $\vec{n}$, which is a unit vector in the ℓ^- rest frame, the differential decay spectrum is of the form [8]:

$$\frac{d\Gamma(\hat{s},\vec{n})}{d\hat{s}} = \frac{1}{2}\left(\frac{d\Gamma(\hat{s})}{d\hat{s}}\right)_0 \left[1 + (P_L e_L + P_T e_T + P_N e_N)\cdot \vec{n}\right], \tag{3}$$

The anti-particle decay rate can be achieved from particle decay rate by the following replacement:

$$\lambda_{tt'} \to \lambda^*_{tt'}, \text{ where } \lambda_{tt'} = \frac{\lambda_{t'}}{\lambda_t} \tag{4}$$

The three different polarization asymmetries calculated in [8].

The $A^i_{CP}(\hat{s})$ terms describe the modification to the unpolarized decay width, which can be written as[8]

$$A^i_{CP}(\hat{s}) = \frac{-4\text{Im}(\lambda_{tt'})\Sigma^i(\hat{s})}{\Delta(\hat{s}) + \Delta(\hat{s})} \approx -2\text{Im}(\lambda_{tt'})\frac{\Sigma^i(\hat{s})}{\Delta(\hat{s})} . \tag{5}$$

where, the explicit expressions for $\Sigma^i(\hat{s})$, $(i = L, N, T)$, are given in [8]It is interesting to note that the polarized CP asymmetries have different combinations involving the imaginary and real parts of C_{10}^{tot} which doesn't appear in unpolarized CP asymmetry.

Longitudinally polarized FB asymmetry for the exclusive decay $B \to K^* \ell^+ \ell^-$ is as follows (see Aliev*et al.*[3]):

$$A^{LL}_{FB} = 16 m_B^4 \hat{s} \frac{v\sqrt{\lambda}}{\Delta} Re[(B_0 B_1^* - C_1 D_1^*)]. \tag{6}$$

Other components are negligible either in the SM3 or SM4 scenario.

NUMERICAL ANALYSIS

We try to analyze the dependency of the asymmetries on the fourth quark mass($m_{t'}$) and the product of quark mixing matrix elements ($V^*_{t'b} V_{t's} = r_{sb} e^{i\phi_{sb}}$).

Using the experimental values of $B \to X_s \gamma$ and $B \to X_s \ell^+ \ell^-$, the bound on $r_{sb} \sim \{0.01 - 0.03\}$ has been obtained for $\phi_{sb} \sim \{0 - 2\pi\}$ and $m_{t'} \sim \{200, 600\}$ (GeV)(see

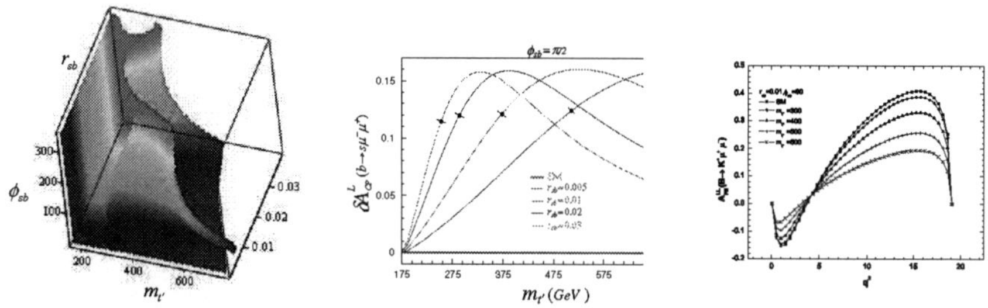

FIGURE 1. Fig. (a) Fig. (b) Fig. (c)

fig. (a)). Considering the B_s mixing, which is in terms of the Δm_{B_s}, ϕ_{sb} receives sharp constrain($\phi_{sb} \sim \pi/2$). Fig. (a) shows the allowed region for the 4th generation parameters considering the 1σ deviation from the measured rate of the $B \to X_s \gamma$ and $B \to X_s \ell^+ \ell^-$. Fig. (b) devoted to the dependency of the averaged longitudinally polarized CP asymmetry on the SM4 parameters. Figs. (c) depict the $\hat{s}$ dependency of FB asymmetry.

To sum up, firstly, we obtain the allowed range for the new parameters. Then, we analyze the polarized CP and FB asymmetries in the allowed region. Both the CP and FB depict strong dependency on the quark mixing matrix product and the 4th generation quark mass. We obtain that Polarized FB coincide with unpolarized FB. Moreover, zero point position of FB depict weak dependency to the new parameters.

REFERENCES

1. S. Eidelman et al., *Phys. Lett.* **B592** (2004) 1.
2. P.H. Frampton, P.Q. Hung and M. Sher, *Phys. Rept.* **330**, 263 (2000).
3. A. Arhrib and W. S. Hou, *Eur. Phys. J.* **C27** (2003) 555; V. Bashiry and F. Falahati, arXiv:0707.3242; T. M. Aliev, V. Bashiry and M. Savci, JHEP **0405**, 037 (2004);Wu-Jun Hou, Tai-Fu Feng, hep-ph/0203207; W. S. Hou, H-n. Li, S. Mishima and M. Nagashima, arXiv:hep-ph/0611107; W. S. Hou, M. Nagashima and A. Soddu, *Phys. Rev. Lett.* **95** (2005) 141601; F. Zolfagharpour and V. Bashiry, arXiv:0707.4337 ; V. Bashiry and K. Azizi, JHEP **0707**, 064 (2007).
4. H.J. He, N. Polonsky and S.F. Su, *Phys. Rev.* **D**, (2001) 052004; V.A. Novikov, L.B. Okun, A.N. Rozanov and M.I. Vysotsky, *Phys. Lett.* **B476**(2000) 107; K. Belotsky, D. Fargion, M. Khlopov, R. Konoplich, K. Shibaev, *Phys. Rev.* **D68** (2003)054027; K. Belotsky, D. Fargion, M.Yu. Khlopov, R.V. Konoplich, M.G. Ryskin, K.I. Shibaev, hep-ph/0411271.
5. E. Arik, S. Atağ, Z. Z. Aydin, A. Çelikel, Z. Çiçek, A. K. Çiftçi, A. Mailov, S. Sultansoy, and G. Ünel, *Phys. Rev.* **D58**, 117701 (1998).
6. G. Buchalla, A. J. Buras and M. E. Lautenbacher, *Rev. Mod. Phys.* **68** (1996) 1125-1144; B. Grinstein, M. J. Savage and M. B. Wise, *Nucl. Phys.* **B319** (1989) 271; G. Cella, G. Ricciardi and A. Viceré, *Phys. Lett.* **B258** 1991 212.
7. A. Ali, G. Hiller, L. T. Handako, and T. Morozomi,*Phys. Rev.* **D55** (1997)4105. (2006).
8. F. Krüger and L. M. Sehgal, *Phys. Rev.* **D55**;K. S. Babu, K. R. S. Balaji, I. Schienbein, *Phys. Rev.* **D68** (1997). (2003).

Damping of Ultrasoft Fermions in Finite Temperature QED

K. Bouakaz* and A. Abada†,**

*Département de Physique, Ecole Normale Supérieure, BP 92 Vieux Kouba, 16050 Alger, Algeria
†Physics Department, United Arab Emirates University, POB 17551, Al Ain, United Arab Emirates
**On leave from Département de Physique, ENS-Kouba, BP 92 Vieux Kouba, 16050 Alger, Algeria

Abstract. We calculate the fermion damping rates to second order in powers of the external momentum in the context of QED at finite temperature using the hard-thermal-loop summation scheme. We find the coefficients of zeroth and first orders finite whereas that of second order logarithmically infrared sensitive. The calculation is done in covariant gauge and the result is independent of gauge fixing.

Keywords: Infrared sensitivity. Fermion damping. Finite-T QED.
PACS: 11.10.Wx 12.20.-m

Perturbative methods in thermal gauge field theories [1, 2] remain of relevance when it comes to describe the physics of the quark gluon plasma formed in high-energy ion collisions. Indeed, above the critical deconfinement temperature $T_c \sim 170$MeV and in the temperature range between T_c and $2T_c$, RHIC results indicate that quarks and gluons seem to deconfine from conventional hadronic states into what it is believed now to be a strongly interacting quark-gluon plasma (sQGP) [3]. This is supported by lattice simulations in which the quark separation energy is a decreasing function of the temperature, starting at about 4GeV when slightly above T_c and reaching 1GeV at about $1.2T_c$ [4]. This is roughly one order of magnitude higher than the temperature itself, which means that perturbative treatment based on quasi-free quarks and gluons is inappropriate in this range of temperatures. But starting from about $3T_c$, weak coupling between individual constituents may be expected. This is in part supported by the behavior of particle susceptibilities: baryons and other bound states contribute in the intermediary temperature range but only quarks survive in the high-temperature limit [5]. This weakly interacting quark-gluon plasma (wQGP) is expected to be seen at the LHC.

QCD perturbative calculations based on bare propagators and vertices lead to gauge dependent and infrared sensitive results [6]. This problem has been partially solved within the hard thermal loop (HTL) summation technique developed in [7] and which yields a finite, positive and gauge-independent zero-momentum transverse-gluon damping rate [8]. A similar computation with similar result for the quarks has also been carried out [9]. However, difficulties still persist at next-to-leading order after HTL reorganization of the perturbation. For example, infrared logarithmic sensitivity is encountered in an early estimation of the damping rate for a heavy fermion [10], and more generally for fast-moving particles [11] and slow-moving quasi-particles [12]. The investigation of the infrared behavior of the quasi-particle dispersion relations in next-to-leading order

CP1006, *High Energy Physics and Applications*, edited by J. Ellis, S. Nasri, and E. Malkawi

HTL perturbation has been furthered in [13] for QCD and in [14] for scalar QED. In this communication, we extend the work to QED: we calculate the damping rates of ultrasoft fermions at next-to-leading order in finite temperature QED. This is a short version of [15].

We use the imaginary-time formalism with Euclidean momenta $P^\mu = (p_0, \mathbf{p})$ so that $P^2 = p_0^2 + p^2$ and $p = |\mathbf{p}|$. We follow the conventions of [2]. The dressed fermion propagator is ${}^*\Delta_f(P) = [\not{P} - \delta\Sigma(P)]^{-1}$ where $\delta\Sigma$ is the HTL in the fermion self-energy. It can be decomposed into the two helicity eigenstates as:

$$
{}^*\Delta_f(P) = -\frac{1}{2}\quad \gamma^0 - i\gamma.\widehat{\mathbf{p}}\ \Delta_+(P) + \ \gamma^0 + i\gamma.\widehat{\mathbf{p}}\ \Delta_-(P)\ , \tag{1}
$$

with the relations:

$$
\Delta_\pm(P) = ip_0 \mp p \mp \frac{m_f^2}{p}\quad 1 - \frac{1}{2}\quad 1 \pm \frac{ip_0}{p}\quad \ln\frac{ip_0 + p}{ip_0 - p}\quad . \tag{2}
$$

At lowest order, the fermion thermal mass $m_f = eT/\sqrt{8}$ where e is the perturbative coupling constant and T the temperature. γ^μ are the euclidean Dirac matrices. The poles of $\Delta_\pm(-i\omega, \mathbf{p})$ give the dispersion relations $\omega_\pm(p)$ for the fermions to lowest order in e. In Covariant gauge, the dressed photon propagator is:

$$
{}^*\Delta_{\mu\nu} = \frac{1}{G+K^2}P^T_{\mu\nu} + \frac{1}{F+K^2}P^L_{\mu\nu} + \frac{\xi}{K^2}\frac{K_\mu K_\nu}{K^2}, \tag{3}
$$

where ξ is a gauge parameter, $P^t_{\mu\nu}$ and $P^l_{\mu\nu}$ the three-dimensional transverse and longitudinal projectors respectively and:

$$
G(K) = m^2 \frac{ik_0}{k}\quad \frac{K^2}{k^2}Q_0\quad \frac{ik_0}{k}\quad + \frac{ik_0}{k}\quad ; \quad F(K) = 2m^2\frac{K^2}{k^2}\quad 1 - \frac{ik_0}{k}Q_0\quad \frac{ik_0}{k}\quad , \tag{4}
$$

with $m = eT/\sqrt{6}$ the photon thermal mass at lowest order and $Q_0(x) = \frac{1}{2}\ln\frac{x+1}{x-1}$. Finally, the dressed fermion-photon vertex and the dressed two-fermion-two-photon vertex, which is a pure hard thermal loop, are given by:

$$
{}^*\Gamma^\mu(P_1, P_2) = \gamma^\mu - m_f^2 \int \frac{d\Omega_s}{4\pi}\frac{S^\mu \not{S}}{P_1 S\, P_2 S}, \tag{5}
$$

$$
{}^*\tilde{\Gamma}^{\mu\nu}(P_1, P_2; K_1, K_2) = -m_f^2 \int \frac{d\Omega_s}{4\pi}\quad \frac{1}{P_1 S} - \frac{1}{P_2 S}\quad \frac{S^\mu S^\nu \not{S}}{(P_1+K_1)S\,(P_2+K_2)S}, \tag{6}
$$

where $S^\mu \equiv (i, \hat{\mathbf{s}})$ and Ω_s is the solid angle of the unit vector $\hat{\mathbf{s}}$.

The next-to-leading-order dressed fermion self-energy is the sum of the two diagrams in figure 1 and is given in imaginary-time formalism by the expression:

$$
{}^*\Sigma(P) = -e^2 \mathrm{Tr}\ {}^*\Gamma^\mu\,(P, Q)\,{}^*\Delta_F(Q)\,{}^*\Gamma^\nu(P, Q)\,{}^*\Delta_{\mu\nu}(K) + {}^*\Gamma^{\mu\nu}(P, -P; Q, -Q)\,{}^*\Delta_{\mu\nu}(K)\ , \tag{7}
$$

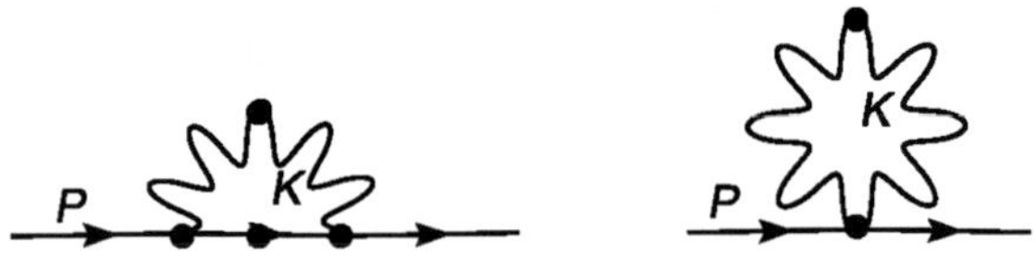

FIGURE 1. Dressed Fermion Self-Energy

where K is the internal soft loop momentum, $Q = P - K$ and $\mathrm{Tr} \equiv T \sum \int \frac{d^3k}{(2\pi)^3}$. In this latter sum, only soft values of k are allowed in the integrals. ${}^*\Sigma(P)$ decomposes into helicity eingenstates in the following form:

$$ {}^*\Sigma(P) = \frac{1}{2} \quad \gamma^0 + i\gamma.\widehat{\mathbf{p}} \quad {}^*f_-(P) + \quad \gamma^0 - i\gamma.\widehat{\mathbf{p}} \quad {}^*f_+(P) \quad , \tag{8} $$

and we can show that at next-to-leading order in the coupling, the damping rates for fermions with ultrasoft momentum p are given by:

$$ \gamma_\pm(p) = \frac{1}{2} \quad 1 \pm \frac{2}{3}\bar{p} - \frac{2}{9}\bar{p}^2 + \ldots \quad \mathrm{Im}\,{}^*f_\pm(-i\omega, p) \Big|_{\omega = \omega_\pm + i0^+} \quad . \tag{9} $$

where we have set $\bar{p} = p/m_f$. Therefore, the calculation of $\gamma_\pm(p)$ amounts to that of the imaginary part of the next-to-leading-order fermion self-energy. Details on how to handle these is given in [15]. At second order in p, we find that the damping rates can be put in the following form:

$$ \gamma_\pm(p) = -\frac{e^2 T}{8\pi} \quad a_0 \pm a_1 \frac{\bar{p}}{3} + a_2 \frac{\bar{p}^2}{9} + \ldots \quad , \tag{10} $$

with:

$$ \begin{aligned} a_i \quad = \quad & \int_\eta^\infty dk \int_{-\infty}^{+\infty} d\omega \int_{-\infty}^{+\infty} \frac{d\omega'}{\omega'} \quad f_i \quad \omega, \omega'; k \quad - f_{i-1, i\geq 1} \quad \omega, \omega'; k \quad \partial_\omega \\ & + f_{i-2, i\geq 2} \quad \omega, \omega'; k \quad -3\,\partial_\omega + \partial_\omega^2 \quad \delta \quad 1 - \omega - \omega' \quad . \end{aligned} \tag{11} $$

In the above expression, η is an infrared cutoff. ∂_ω stands for $\partial/\partial\omega$ and ∂_ω^2 for the second derivative. The integrands $f_i(\omega, \omega'; k)$ are generically polynomials in ω multiplied by the product of two spectral distributions or their derivatives. The coefficients of the polynomials in ω are sums of positive and negative powers of k. These three functions are long and will not be displayed [15].

The integrals in the coefficients a_0, a_1 and a_2 are evaluated numerically after the extraction of all infrared-sensitive contributions. In fact, during the calculation, we do encounter two other kinds of divergences: light-cone and at specific points defined by the photon dispersion relations. All potential divergences can be circumvented except the infrared ones. We find the final gauge-independent result:

$$ \gamma_\pm(p) = \frac{e^2 T}{16\pi} \quad 5.4252 \mp 0.64736\,\bar{p} + (0.88258 - 13.199 \ln \bar{\eta})\,\bar{p}^2 + \mathcal{O} \quad \bar{p}^3 \quad . \tag{12} $$

As can been seen, the first and second coefficients are finite but the third one is logarithmically infrared sensitive.

ACKNOWLEDGMENTS

This work was done in collaboration with D. Deguiche.

REFERENCES

1. N. P. Landsman and Gh. G. Vanweert, Phys. Rep. **145** (1987) 141; J. I. Kapusta, '*Finite-temperature field theory*', Cambridge University Press, 1989.
2. M. Le Bellac, '*Thermal Field Theory*', Cambridge University Press, 1996.
3. E. Shuryak, Quark Matter 05, Budapest, August 2005, hep-ph/0510123; E. Shuryak and I. Zahed, Phys. Rev. C**70** (2004) 021901, Phys. Rev. D**69** (2004) 014011; Prog. Part. Nucl. Phys. **53** (2004) 273.
4. O. Kaczmarek, S. Ejiri, F. Karsch, E. Laermann and F. Zantow, Prog. Theor. Phys. Suppl. 153 (2004) 287.
5. J. Liao and E. Shuryak, hep-ph/0510110.
6. V.P. Silin, Zh. Eksp. Teor. Fiz. **38** (1960) 1577 [Sov. Phys. JETP **11** (1960) 1136]; V.N. Tsytovich, Zh. Eksp. Teor. Fiz. **40** (1961) 1775 [Sov. Phys. JETP **13** (1961) 1249]; O.K. Kalashnikov and V.V. Klimov, Yad. Fiz. **31** (1980) 1357 [Sov. J. Nucl. Phys. **31** (1980) 699]; V.V. Klimov, Yad. Fiz. **33** (1981) 1734 [Sov. J. Nucl. Phys. **33** (1981) 934], Zh. Eskp. Teor. Fiz. **82** (1982) 336 [Sov. Phys. JETP **55** (1982) 199]; H.A. Weldon, Phys. Rev. D**26** (1982) 1394.
7. E. Braaten and R.D. Pisarski, Nucl. Phys. **B**337 (1990) 569, Phys. Rev. Lett. 64 (1990) 1338, Nucl. Phys. **B**339 (1990) 310; R.D. Pisarski, Phys. Rev. Lett. 63 (1989) 1129, R.D. Pisarski, Physica **A** 158 (1989) 246, Nucl. Phys. **A**498 (1989) 423c; J. Frenkel and J.C. Taylor, Nucl. Phys. **B**334 (1990) 199.
8. E. Braaten and R.D. Pisarski, Phys. Rev. **D**42 (1990) R2156.
9. R. Kobes, G. Kunstatter and K. Mak, Phys. Rev. **D**45 (1992) 4632; E. Braaten and R. Pisarski, Phys. Rev. **D**46 (1992) 1829.
10. R.D. Pisarski, Phys. Rev. Lett. 63 (1989) 1129.
11. V. V. Lebedev and A. V. Smilga, Ann. Phys. **202** (1990) 229, Phys. Lett. **B**253 (1991) 231; C. P. Burgess and A. L. Marini, Phys. Rev. **D**45 (1992) 17; A. Rebhan, Phys. Rev. **D**46 (1992) 482.
12. R. Pisarski, Phys. Rev. **D**47 (1993) 5589; F. Flechsig, A. Rebhan and H. Schulz, Phys. Rev. **D**52 (1995) 2994.
13. A. Abada, K. Bouakaz and O. Azi, Phys. Scr. **74** (2006) 77; A. Abada, N. Daira-Aifa and K. Bouakaz, Int. Jour. Mod Phys. **A** 21 (2006) 5317; A. Abada, K. Bouakaz, N. Daira-Aifa, Eur. Phys. J. **C**18 (2001) 765; A. Abada, et O. Azi, Phys. Lett. **B**463 (1999) 117; A. Abada, O. Azi and K. Benchallal, Phys. Lett. **B**425 (1998) 158.
14. A. Abada and K. Bouakaz, JHEP01 (2006) 161.
15. A. Abada, K. Bouakaz and D. Deghiche, Mod. Phys. Lett. **A**22 (2007) 903.
16. R. D. Pisarski, Physica **A** 158 (1989) 146; Nucl. Phys. **B**309 (1988) 476.

Toward a Strongly Interacting Scalar Higgs Particle

Abouzeid M. Shalaby* and M. El-Houssieny†

*Physics Department, Faculty of Science, Qassim University, KSA[1]
†Mathematics Department, Faculty of Science, Al-Azhar University, Egypt

Abstract. We calculate the vacuum energy of the non-Hermitian and PT symmetric $\left(-g\phi^4\right)_{2+1}$ scalar field theory. Rather than the corresponding Hermitian theory and due to the asymptotic freedom property of the theory, the vacuum energy does not blow up for large energy scales which is a good sign to solve the hierarchy problem when using this model to break the $U(1) \times SU(2)$ symmetry in the standard model. The theory is strongly interacting and in fact, all the dimensionful parameters in the theory like mass and energy are finite even for very high energy scales. Moreover, relative to the vacuum energy for the Hermitian ϕ^4 theory, the vacuum energy of the non-Hermitian and PT symmetric $\left(-g\phi^4\right)_{2+1}$ theory is tiny, which is a good sign toward the solution of the cosmological constant problem. Remarkably, these features of the non-Hermitian and PT symmetric $\left(-g\phi^4\right)_{2+1}$ scalar field theory make it very plausible to be employed as a Higgs mechanism in the standard model instead of the problematic Hermitian Higgs mechanism.

Keywords: effective potential, non-Hermitian models, PT symmetric theories.
PACS: 11.10.Kk, 02.30.Mv, 11.10.Lm, 11.30.Er, 11.30.Qc, 11.15.Tk

One of the most challenging problems in the conventional standard model are the Hierarchy and triviality problems. The Hierarchy problem has its manifestations in all the parameters of the theory which has a positive mass dimension. In fact, this problem arising when we try to make these parameters to flow as the energy scale goes to very high values like the Planck scale. This flow is needed, because if the Higgs particle exists as a fundamental particle it should be a part of any unified theory and the theory that describe it (ϕ^4) has to be free of problems at the grand unification scale. However, it is very clear that the Higgs mechanism used in the conventional standard model is suffering from two genuine problems one for high energy scales (Hierarchy) and one for small energy scale (triviality).

In 1970 Kurt Symanzik proposed a 'precarious' ϕ^4-theory with a negative quartic coupling constant as a valid candidate for an asymptotically free theory of strong interactions [1, 2, 3, 4]. At that time, no body was ready to accept this theory for the wrong believe that it is not a stable theory. However, in 1998, Carl Bender and Stefan Boettcher have shown that a class of Non-Hermitian but PT symmetric Hamiltonians have real and bounded from below spectra [5]. This discovery led us to reinvestigate the non-Hermitian quantum field models like the $(-g\phi^4)_{1+1}$ and $\left(-g\phi^4+h\phi^6\right)_{1+1}$[2] scalar theories [8, 7]. In the first model, we realized that, rather than the corresponding Hermi-

[1] Permanent address: Physics Department, Faculty of Science, Mansoura University, Egypt.
[2] There exists One phase for which the theory is non-Hermitian.

CP1006, *High Energy Physics and Applications*, edited by J. Ellis, S. Nasri, and E. Malkawi

tian model $(g\phi^4)$, the vacuum energy is small for the full range of energy scale (except for the IR slavery region). Besides, the vacuum energy is real and in fact, it is easy to show that the *PT* symmetry is exact which verifies that not only the lowest energy but the full energy spectrum is real [6]. Moreover, we discovered that the Hermitian model $(g\phi^4 + h\phi^6)$ can have a transition to a phase for which the theory is physically acceptable though non-Hermitian [7]. In fact, we asserted that this model has a very interesting property namely, matter-antimatter asymmetry, which for the first time was obtained in a spontaneously symmetry breaking regime. Relying on such interesting properties of PT symmetric and non-Hermitian models, one may ask if such new subject can help in solving existing problems in Physics like the cosmological constant problem. In this, work we investigate the vacuum energy of the non-Hermitian and *PT* symmetric $(\phi^4)_{2+1}$, which, due to the asymptotic freedom property, is expected to be tiny compared to that of the corresponding Hermitian theory.

To start, consider a Hamiltonian that is normal ordered with respect to the vacuum of mass parameter m;

$$H = N_m\left(\frac{1}{2}\left((\nabla\phi)^2 + \pi^2 + m^2\phi^2\right) - \frac{g}{4}\phi^4\right), \tag{1}$$

After simple algebraic steps, we can get the Hamiltonian form [8];

$$H = \bar{H}_0 + \bar{H}_I + \bar{H}_1 + E, \tag{2}$$

where

$$\bar{H}_0 = N_M\left(\frac{1}{2}\left(\Pi^2 + (\psi)^2\right)\right) + \frac{1}{2}N_M\left(m^2 - 3g\left(B^2 + \Delta\right)\right)\psi^2, \tag{3}$$

and

$$\bar{H}_I = \frac{-g}{4}N_M\left(\psi^4 + 4B\psi^3\right). \tag{4}$$

$\bar{H}_1$ can be found as

$$\bar{H}_1 = N_M\left(m^2 - g\left(B^2 + 3\Delta\right)\right)B\psi, \tag{5}$$

and the field independent terms can be regrouped as

$$E = \frac{1}{2}\left(m^2 - \frac{12g\Delta}{4}\right)B^2 - \frac{g}{4}B^4 + \frac{1}{8\pi}\left(M^2 - m^2\right) - \frac{3g\Delta^2}{4} + \frac{1}{2}m^2\Delta. \tag{6}$$

To understand well the features of this result, let us note that the renormalization conditions are given by [9]

$$\frac{\partial^n}{\partial b^n}E(b,t,G) = g_n, \tag{7}$$

where g_n is the ψ^n coupling. For instance,

$$\frac{\partial^2 E}{\partial B^2} = M^2, \tag{8}$$

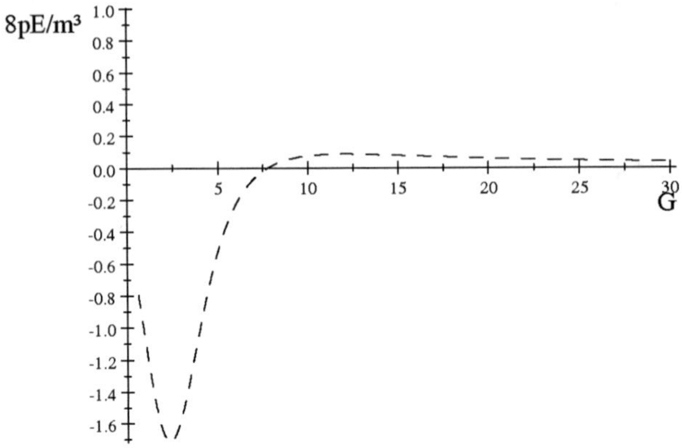

FIGURE 1. The vacuum energy as a function of the the coupling G.

where $g_2 = M^2 = m^2 - 3g\left(B^2 + \Delta\right)$. Note that, the renormalization condition $\frac{\partial E}{\partial B} = 0$ enforces $\overline{H}_1$ to be zero.

Since normal ordering is a first order calculation in the coupling as well as it accounts for all the cactus diagrams, going to higher orders one need to include non-cactus diagrams only. Accordingly, up to two loops, we include the sunset diagram. Hence, the vacuum energy as well as the renormalization condition equations take the form;

$$0 = \left\{1 - \frac{G}{6}\left(3\left(1-t\right) + b^2\right) + \frac{G^2}{6}\ln t\right\} b, \tag{9}$$

$$t^2 = \left\{1 - \frac{G}{2}\left(\left(1-t\right) + b^2\right) + \frac{G^2}{6}\ln t\right\}, \tag{10}$$

$$\frac{8\pi E}{m^3} = \left\{ \begin{array}{c} b^2 - \frac{G}{12}b^4 - \frac{1}{2}b^2 G\left(1-t\right) + \frac{G^2}{6}\ln t\left(b^2 + \left(1-t\right)\right) \\ -\frac{G}{4}\left(1-t\right)^2 + \frac{1}{3}\left(t^3 - 3t + 2\right) \end{array} \right\}, \tag{11}$$

where $G = \frac{g}{4\pi m}$, $t = \frac{M}{m}$ and $b = B\sqrt{\frac{4\pi}{m}}$. To get the vacuum parameters b and t, we solve Eqs.(9,10) and substitute in Eq.(11) to get the vacuum energy. As we can see from Fig.1, the vacuum energy is tiny for all the coupling range and it goes to zero for very high energy scales. To understand this behavior, consider the β function of the theory , which, up to five loops, is given by [10]

$$\beta\left(g\right) = -\left(1.5g^2 + 2.833g^3 + 16.27g^4 + 135.8g^5 + 1424.2841g^6\right), \tag{12}$$

which is always negative and thus the theory is asymptotically free. Accordingly, at high energy scales, the vacuum energy tends to zero which agrees with our numerical calculations.

To conclude, we calculated the vacuum energy of the non-Hermitian and PT symmetric $\left(-g\phi^4\right)_{2+1}$ scalar field theory. Due to the asymptotic freedom property of the theory, the vacuum energy does not blow up for large energy scales which is a good sign to solve the hierarchy problem when using this model to break the $U(1) \times SU(2)$ symmetry in the standard model.

ACKNOWLEDGMENTS

One of the authors would like to thank Dr. Suleiman Al-Thoyaib for his support and kind help.

REFERENCES

1. Frieder Kleefeld, J. Phys. A: Math. Gen. 39 L9–L15 (2006).
2. Symanzik K, Springer Tracts Mod. Phys. 57 222–36 (1971).
3. Symanzik K, Commun. Math. Phys. 23 49–86, (1971).
4. Symanzik K, Nuovo Cimento 6 77–80, (1973).
5. Carl Bender and Stefan Boettcher, Phys.Rev.Lett.80:5243-5246 (1998).
6. Abouzeid shalaby, Will the *PT* Symmetric and Non-Hermitian ϕ^4 Theory Solve the Hierarchy and Triviality Problems in the Standard Model? hep-th/0712.2521.
7. Abouzeid shalaby, Phys.Rev.D76:041702 (2007).
8. Abouzeid shalaby, Eur.Phys.J.C50:999-1006 (2007).
9. Michael E.Peskin and Daniel V.Schroeder, AN INTRODUCTION TO THE QUANTUM FIELD THEORY (Addison-Wesley Advanced Book Program, 1995).
10. H. Kleinert, J. Neu, V. Schulte-Frohlinde, K.G. Chetyrkin, S.A. Larin, Phys.Lett.B272:39-44 (1991), Erratum-ibid.B319:545 (1993).

POSTER SESSION

Quark-Gluon Plasma Model and Origin of Magic Numbers

N. Ghahramany, M. Ghanaatian and M. Hooshmand

Dept. of Physics, Shiraz University, Shiraz 71454, Iran

Abstract. Using Boltzman distribution in a quark-gluon plasma sample it is possible to obtain all existing magic numbers and their extensions without applying the spin and spin-orbit couplings. In this model it is assumed that in a quark-gluon thermodynamic plasma, quarks have no interactions and they are trying to form nucleons. Considering a lattice for a central quark and the surrounding quarks, using a statistical approach to find the maximum number of microstates, the origin of magic numbers is explained and a new magic number is obtained.

Keywords: Magic number, Quark-Gluon Plasma
PACS: 21.10.-k, 25.75.Nq, 21.60.Cs, 12.38.Mh, 21.65.Qr

INTRODUCTION

There are certain elements in the universe with relative high stability and abundance whose neutron or proton numbers are called magic numbers. Historically, their stability and excited energies were first found [1] but the origin of magic numbers remained as a mystery. Of course there are some explanations about these numbers from shell model of nuclei, mainly from observation of apparent similarities between these magic numbers and nucleon numbers that fill the nuclear shells. In nuclear shell model it is assumed that the nucleon is orbiting in a nuclear spherical potential well and the energy gaps between the spectral lines obtained from such potential well correspond to the stability of nuclei. Such correspondence was not accurate enough, therefore several researchers, mainly Maria Goepert Mayer [2] included the effect of spin and spin-orbit coupling in the nuclear Hamiltonian as a perturbation from which new energy gaps were observed in more agreement with the observed magic numbers. Eventually the shell model was built explaining the nuclear structure and position of constituent particles with no satisfactory explanation about the origin of the magic numbers.

In this research it is intended to investigate the origin of these magic numbers via quark-gluon plasma media. In this model it is assumed that in a quark-gluon thermodynamic plasma in which the quark have no interaction, quark are trying to form nucleons. If we accept that the stability of a thermodynamical system is obtained when the system is in maximum disorder or maximum number of complexions, then by considering different isolated system containing one central quark, and 2, 3, 4, 5, 6, 7 and at most 8 surrounding quarks embracing the central quark, these can find maximum number of microstates that correspond exactly to magic numbers. From statistical point of view it will be clear that why it is unlikely to have a central quark with 8 surrounding quarks and therefore very likely to have a central one with 2 or 3 quarks around it.

CP1006, *High Energy Physics and Applications*, edited by J. Ellis, S. Nasri, and E. Malkawi

COLORED QUARK-GLUON PLASMA AND MAGIC NUMBERS

The hot quark-gluon plasma (QGP) exists right after big-bang and by relativistic expansion cool down and change to proton and neutron. In the continued process of expansion, different nuclei are formed (nucleation) via Boltzman equilibrium process [3-7]. These formed nuclei are most stable at magic numbers. We intend to investigate how and under what conditions the quarks with color and flavor form proton and neutron and what the origin of magic numbers is in such QGP. It is not intended to describe the quark distribution after the formation of proton which is given in terms of structure functions.

If the QGP is considered as a thermodynamical media then it should proceed toward maximum disorder. It should be investigated that how such system approach equilibrium. The thermodynamical state is a stable system with maximum probability state, i.e., the most probable state with maximum number of complexion.

Now consider a thermodynamical state of quarks in motion. In such QGP soup the quarks are not absolutely free. This is known from lattice QCD theory [8]. In fact in such QGP soup, the gluons connect to the nearby quarks with a force much weaker than the binding force. The QGP media is assumed as an ideal gas model. In such model consider a quark to be trying to form a nucleon capturing two quarks of different flavor. In such competing space between quarks different nucleon formation cases happen. In Fig. 1 quarks are considered in a lattice media and the central u-quark seeks 2 d-quarks to form a proton.

If there is only two d-quark with no color neighboring the central u-quark, then there is no competition and state ud_1d_2 is formed. But from standard model each quark have 3 color and d_1 and d_2 must be of different color say green and blue, therefore two competing cases exist namely, a red u with blue d_1 and green d_2 or with green d_1 and blue d_2. So we get number 2 as the first magic number.

Now consider that there are three d-quarks neighboring the central u-quark, then there are three cases namely, ud_1d_2, ud_1d_3 and ud_2d_3. If their color is also taken into account, then there are six cases in addition to the previous two cases and we get eight cases to form a proton, i.e., the second magic number. Lets consider four neighboring d-quarks. If only two of them compete, we have 2 cases and if three of them participate in this competition then we have six cases and if all four compete then we get 12 cases. The total number of cases is therefore 2+6+12=20.This is the third magic number.

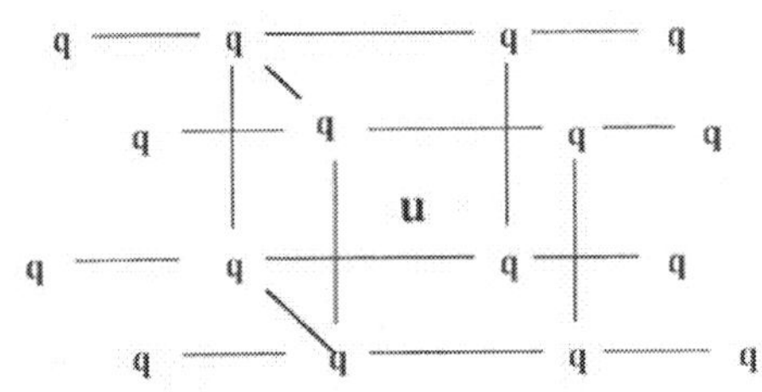

FIGURE 1. Cubic quark-gluon plasma lattice with central u-quark.

It is interesting to note that in both the spin and spin-orbit coupling interpretations to explain magic numbers [9] and numerical explanation of Bagge [10], two separate series were introduced namely,

$$\begin{aligned} M_1 &\in \underline{2,\ 8,\ 20},\ 40,\ 70,\ 112 \\ M_2 &\in 2,\ 6,\ 14,\ \underline{28,\ 50,\ 82,\ 126} \end{aligned} \qquad (1)$$

After the magic number 20, there was a jump from the first series to the second series to obtain number 28.

If we consider 5 quarks neighboring the central one, the competition between these 5 quarks in addition to previous ones given us 40 cases which is exactly the fourth number from the first series and for 6 and 7 neighboring quarks 70 and 112 cases are obtained and the same historical problem do exist. To resolve this problem in our model, the QGP condition is utilized and "imposed quarks" are introduced.

From lattice QCD theory it is clear that quarks are not free and there exist some weak attractive forces between quarks in QGP soup and that is why it is called a soup. Now suppose there are four *d* quarks neighboring the central *u*-quark one as obtained before there are 20 cases competing to form a proton. If each *d*-quark is considered to be close to its own neighbors as seen in Fig. 2 then if for example *d* is absorbed by it then the closest quark to *d* which has the strongest attraction force to *d* will accompany it and participate. Lets call it *d'* and this is named as imposed quark.

This quark come from the second level, therefore with one imposed quark for each initial 4 *d*-quarks we have $ud_1d'_1$,$ud_2d'_2$,$ud_3d'_3$ and $ud_4d'_4$ and considering their colors there are 8 cases in addition to the pervious 20 cases, adding to 28 competing states to form a proton.

Lets consider five *d*-quarks surrounding the central *u*-quark, then we have twenty new cases in addition to ten imposed cases adding to 50 cases 20+20+10=50, which consist of ud_1d_2, ud_1d_3, ud_1d_4, ud_1d_5, ud_2d_3, ud_2d_4, ud_2d_5, ud_3d_4, ud_3d_5, ud_4d_5 and $ud_1d'_1$, $ud_2d'_2$, $ud_3d'_3$, $ud_4d'_4$, and $ud_5d'_5$.

Now lets consider six *d*-quarks around the central *u*-quark, then we have: 40+30+12=82. For seven *d*-quarks participating: 70+42+14=126.

Eventually for eight *d*-quarks we get 184 cases. This is a new magic number that is obtained in this model. For more than 8 quarks one has to consider the imposed quarks from the third level and no additional magic number is obtained.

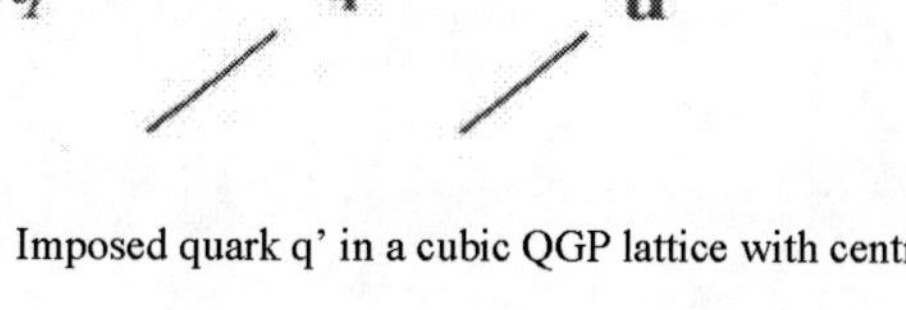

FIGURE 2. Imposed quark q' in a cubic QGP lattice with central *u*-quark.

CONCLUSION

A quark-gluon plasma soup model is presented based upon Boltzman distribution and an alternative approach is suggested to obtain not only the existing magic numbers exactly but new magic number is introduced. A quark shell structure in the form of cubic lattice is considered to find the most probable cases and maximum number of ways that a stable element is formed corresponding to the magic numbers. While work is in progress to understand how exactly the same magic numbers appear in nuclear formation, this paper is intended to provide insight and extended the concept of magic numbers from nuclei to nucleon formation independently.

ACKNOWLEDGMENTS

Partial support of the research council of Shiraz University is appreciated.

REFERENCES

1. W. Elsasser, *J. Phys. Radium* **4**, 549 (1933).
2. M. G. Mayer, *Phys. Rev.* **74**, 235 (1948); M. G. Mayer, *Phys. Rev.* **75**, 1969 (1949); M. G. Mayer, and J. H. D. Jensen, *Elementary Theory of Nuclear Shell Structure*, John Wiley & Sons, New York, 1955.
3. A. Lefebvre et al, *Nuclear Physics* **A621**, 199 (1997).
4. R. Kippenhahn, and A. Weigert, *Stellar Structure and Evolution*, Springer, Heidelberg, 1990.
5. F. Osman, N. Ghahramany, and H. Hora, *Laser and Particle Beams* **23**, 461-466, (2005).
6. H. Hora, *Plasma Model for Surface Tension of Nuclei and the Phase Transition to the Quark Plasma*, Report CERN-PS/DL-Note-91/05, 1991.
7. T. Ranscher, J. H. Applegate, J. J. Cowan, F. K. Thielmann, and M. Weiseher, *Astrophys. J.* **429**, 499 (1994).
8. K.G. Wilson, *Phys. Rev.* **D10**, 2445 (1974); K. G. Wilson, *Rev. Mod. Phys.* **55**, 583 (1983); K. G. Wilson et al, *Phys. Rev.* **D49**, 6720 (1994).
9. O. Haxel, J. H. D. Jensen and H. E. Suess, Zeitscher. *F. Physik* **128**, 295 (1950).
10. E. Bagge, *Naturwissenschaften* **35**, 376 (1948).

Pion production off the deuteron with real photons including polarization observables

E. M. Darwish[1,*], C. Fernández-Ramírez[2,†], E. Moya de Guerra[†,**] and J. M. Udías[†]

*Physics Department, Faculty of Science, Sohag University, Sohag 82524, Egypt
†Grupo de Física Nuclear, Departamento de Física Atómica, Molecular y Nuclear, Facultad de Ciencias Físicas, Universidad Complutense de Madrid, Avda. Complutense s/n, E-28040 Madrid, Spain
**Instituto de Estructura de la Materia, CSIC, Serrano 123, E-28006 Madrid, Spain

Abstract. Recent results are presented for quasifree pion photoproduction off the deuteron in the energy region from threshold up to the $\Delta(1232)$-resonance with inclusion of all *leading* πNN effects. Final-state interaction effects are investigated and their role in polarized and unpolarized observables are found to be significant. The sensitivity of results to the elementary $N(\gamma,\pi)N$ operator is discussed and considerable dependence is found. We expect that these results may be useful to interpret the recent measurements from LEGS, A2 and GDH Collaborations.

Keywords: Photonuclear reactions; meson production; spin observables
PACS: 25.20.Lj; 13.60.Le; 14.20.-c

1. MOTIVATION

Meson photo- and electroproduction on light nuclei is an important topic in medium energy nuclear physics. It is motivated by the following aspects: (i) study of the elementary reaction on the neutron to which otherwise one has no access, (ii) investigation of the influence of a nuclear environment on the elementary production amplitude, and (iii) it provides an interesting mean to study nuclear structure. As an illustration of these various aspects, we present recent results on quasifree pion photoproduction from the deuteron including polarization observables [1]. These spin-dependent observables allow us to obtain more information on the reaction mechanism than unpolarized observables. The latter only provide information on the sum of the absolute squares of the reaction amplitudes.

The calculations presented in this article are of high interest in view of the extensive recent polarization measurements from the A2 and GDH@MAMI and LEGS@BNL Collaborations. A theoretical understanding of these data on the deuteron target will provide information about the pion photoproduction from the neutron reaction, which is not so well known yet, but is required for a complete understanding of Δ-excitations in the pion photoproduction process.

---

[1] Corresponding author; *e-mail address:* darwish@sci.sohag.edu.eg
[2] Present address: Center for Theoretical Physics, Massachusetts Institute of Technology, 77 Massachusetts Ave., Cambridge, MA 02139, USA

CP1006, *High Energy Physics and Applications*, edited by J. Ellis, S. Nasri, and E. Malkawi

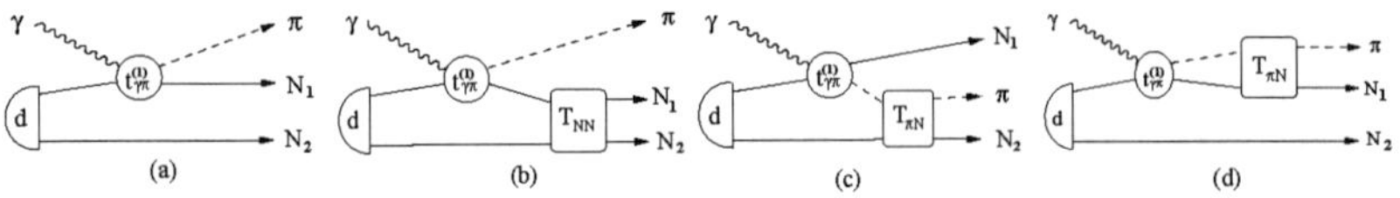

FIGURE 1. Diagrammatic representation of the $\gamma d \rightarrow \pi NN$ amplitude: (a) impulse approximation (IA); (b), (c), and (d) "driving terms" from NN- and πN-rescattering, respectively.

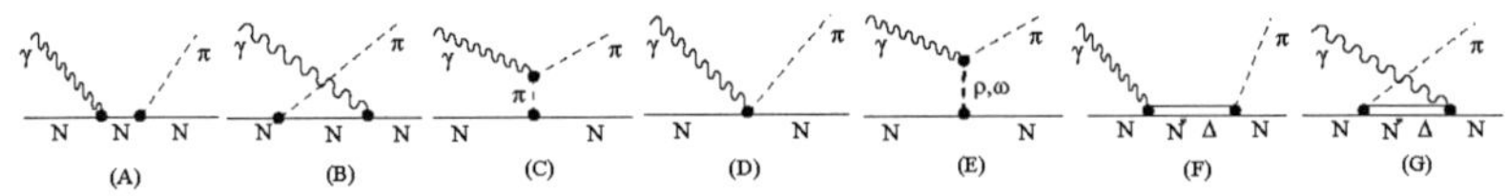

FIGURE 2. Feynman diagrams for the elementary process $\gamma N \rightarrow \pi N$. Born terms: (A) direct, (B) crossed (C) pion in flight, and (D) Kroll-Rudermann; (E) vector-meson exchange (ρ and ω); resonance excitations contribution: (F) direct and (G) crossed.

2. THE MODEL

In [1] we have included, besides the pure impulse approximation (IA), the complete rescattering by the final-state interaction within each of the two-body subsystems NN and πN. Therefore, the total transition matrix element reads

$$T_{sm\,\mu m_d} = T^{IA}_{sm\,\mu m_d} + T^{NN}_{sm\,\mu m_d} + T^{\pi N(c)+(d)}_{sm\,\mu m_d} \,. \tag{1}$$

A graphical representation of the scattering matrix is shown in Fig. 1. The necessary half-off-shell two-body scattering matrices were obtained from separable representations of realistic NN and πN interactions. For the elementary $N(\gamma,\pi)N$ amplitude, we have taken a realistic operator from [2] which displays chiral symmetry, gauge invariance, and crossing symmetry as well as a consistent treatment of the spin-3/2 interaction. The model is based upon an effective Lagrangian approach (ELA) which from a theoretical point of view is an appealing, reliable, and formally well-established approach in the energy region of the mass of the nucleon. The model includes Born terms (diagrams (A)-(D) in Fig. 2), vector-meson exchanges (ρ and ω, diagram (E) in Fig. 2), and all the four star resonances in Particle Data Group [3] up to 1.7 GeV and up to spin-3/2: Δ(1232), N(1440), N(1520), N(1535), Δ(1620), N(1650), and Δ(1700) (diagrams (F) and (G) in Fig. 2). It has been successfully applied to pion photoproduction from the nucleon [2, 4, 5]. The explicit expressions for the various contributions in (1) can be found in [1, 6].

3. RESULTS

Quasifree pion photoproduction from the deuteron in the energy region from threshold up to the Δ(1232)-resonance is investigated [1] where besides the pure impulse approximation (IA) complete two-body rescattering in the NN and πN final state subsystems is included (see Fig. 1). For the elementary $\gamma N \rightarrow \pi N$ amplitude, a realistic ELA from [2] is taken. The necessary half-off-shell two-body scattering matrices were obtained

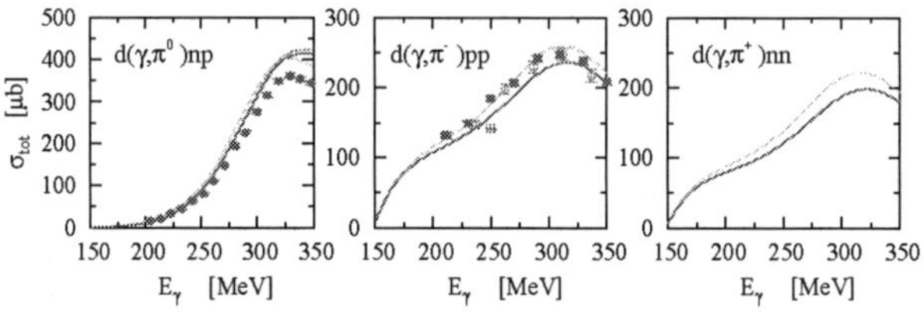

FIGURE 3. Total cross section for $d(\gamma,\pi)NN$ using different elementary pion photoproduction operators and including FSI effects. Solid: IA+$NN+\pi N$ using the bare electromagnetic multipoles of ELA [2]; dashed: IA* $+NN+\pi N_c$ using MAID [7]; dotted: IA* $+NN+\pi N_c$ using the dressed multipoles of ELA. IA* denotes the calculation when πN-FSI is included in the elementary reaction (see [1]). Exp. data: Krusche *et al.* [8] (solid circles) for π^0 channel, Benz *et al.* [9] (solid square), Chiefari *et al.* [10] (open circles) and Quraan *et al.* [11] (open square) for π^- channel.

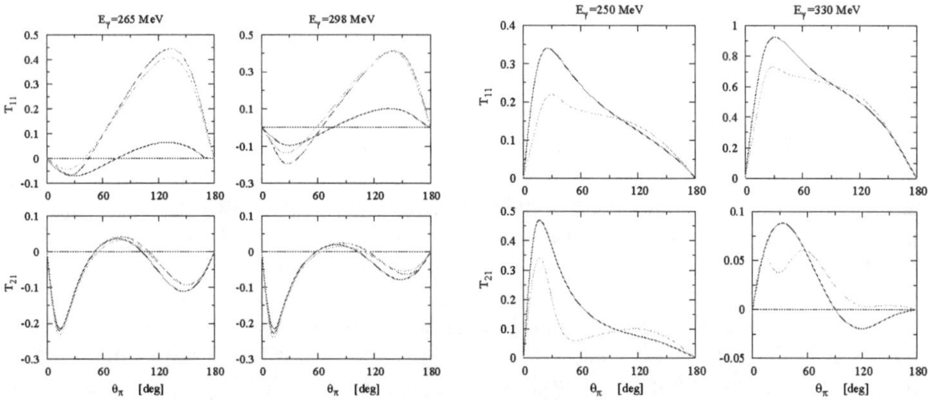

FIGURE 4. Single-spin asymmetries T_{11} and T_{21} using different elementary pion photoproduction operators and including FSI effects. Left group: results for $\gamma\vec{d}\rightarrow\pi^0 np$ at $E_\gamma = 265$ and 298 MeV. Right group: results for $\gamma\vec{d}\rightarrow\pi^- pp$ at $E_\gamma = 250$ and 330 MeV. Curves conventions as in Fig. 3.

from separable representations of realistic NN and πN-interactions which gave good descriptions of the corresponding phase shifts.

The results we present in Fig. 3 show the total cross section for separate channels of $d(\gamma,\pi)NN$ using both the ELA [2] and MAID-2003 [7] models. Fig. 4 shows the results for the single-spin asymmetries T_{11} and T_{21} for both π^0 and π^- channels. In addition, we have evaluated the contribution of separate channels to the Gerasimov-Drell-Hearn (GDH) sum rule [12] as shown in Fig. 5. We found that the inclusion of final state interaction (FSI) effects is important and leads, in many cases, to a good agreement with experimental data. The spin asymmetries are found to be very sensitive to FSI effects. Furthermore, certain observables are found to be sensitive to the elementary operator. In many cases, the deviation among results obtained using different operators is sizeable. For example, the total value $I^{GDH}_{IA+NN+\pi N}(350\text{ MeV}) = 41.29\ \mu$b has been obtained for

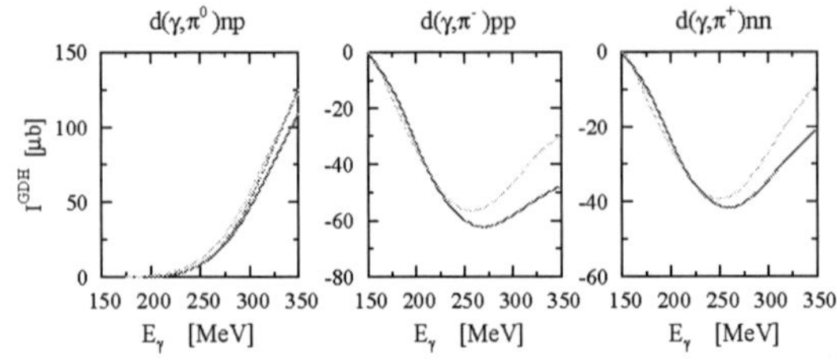

FIGURE 5. The Gerasimov-Drell-Hearn integral as a function of the upper integration limit for separate channels of the $\vec{d}(\vec{\gamma},\pi)NN$ for different operators. Curve conventions as in Fig. 3.

the GDH integral using the ELA model, whereas the value $I^{GDH}_{IA^*+NN+\pi N}(350\text{ MeV}) = 84.61\ \mu\text{b}$ has been obtained using the MAID-2003 model.

In summary, we conclude that certain spin observables of the $d(\gamma,\pi)NN$ process can be used as a filter for different elementary operators.

ACKNOWLEDGMENTS

E.M.D. would like to thank the organizers for their warm hospitality and financial support and A. Abdulnaby for keen interest to provide assistance with computing difficulties. E.M.G. acknowledges partial support from Ministerio de Educación y Ciencia (Spain) under contract FIS2005-00640. J.M.U. acknowledges partial support from Ministerio de Educación y Ciencia (Spain) under contracts FPA2006-07393 and FPA2007-62216. E.M.G. and J.M.U. acknowledge partial support from UCM and Comunidad de Madrid under project number 910059 (Grupo de Física Nuclear). C.F.-R. is supported by "Programa de Becas Posdoctorales" of Ministerio de Educación y Ciencia (Spain).

REFERENCES

1. E. M. Darwish, C. Fernández-Ramírez, E. Moya de Guerra, and J.M. Udías, *Phys. Rev. C* **76**, 044005 (2007); submitted for publication (2007).
2. C. Fernández-Ramírez, E. Moya de Guerra, and J.M. Udías, *Ann. Phys. (N.Y.)* **321**, 1408 (2006).
3. Review of Particle Physics, W.-M. Yao *et al.*, *J. Phys. G* **33**, 1 (2006).
4. C. Fernández-Ramírez, E. Moya de Guerra, and J.M. Udías, *Phys. Rev. C* **73**, 042201(R) (2006); *Eur. Phys. J. A* **31**, 572 (2007).
5. C. Fernández-Ramírez, E. Moya de Guerra, and J.M. Udías, *Phys. Lett. B*, doi:10.1016/j.physletb.2007.11.099 (2008).
6. E.M. Darwish, H. Arenhövel, and M. Schwamb, *Eur. Phys. J. A* **16**, 111 (2003).
7. D. Drechsel, O. Hanstein, S. Kamalov, and L. Tiator, *Nucl. Phys. A* **645**, 145 (1999); *Phys. Rev. D* **63**, 114010 (2001); MAID Program, `http://www.kph.uni-mainz.de/MAID/maid2003/`.
8. B. Krusche, J. Ahrens, R. Beck, M. Fuchs, S.J. Hall, F. Härter, J.D. Kellie, V. Metag, M. Röbig-Landau, and H. Ströher, *Eur. Phys. J. A* **6**, 309 (1999).
9. P. Benz *et al.*, *Nucl. Phys. B* **65**, 158 (1973).
10. G. Chiefari, E. Drago, M. Napolitano, and C. Sciacca, *Lett. Nuovo Cim.* **13**, 129 (1975).
11. M.A. Quraan *et al.*, *Phys. Rev. C* **57**, 2118 (1998).
12. S.B. Gerasimov, *Sov. J. Nucl. Phys.* **2**, 430 (1966) [*Yad. Fiz.* **2**, 598 (1965)]; S.D. Drell and A.C. Hearn, *Phys. Rev. Lett.* **16**, 908 (1966).

The Discrete Super space Formalism and the Quantization of the Standard Model

A.Redouane-Salah[a] and N.Mebarki[a]

[a] *Laboratoire de Physique Mathématique et Subatomique*
Faculty of Science, Mentouri University, Constantine, Algeria

Abstract. Using an extended discrete superspace approach, a mathematical formalism to quantize the Weinberg-Salam-Glashow standard model is presented and the corresponding BRS transformations are derived through an horizontality like condition.

Keywords: : BRST Quantization, Discrete space, Gauge
PACS: 11.10.Nx, 12.60.-I.

INTRODUCTION

The construction of Noncommutative gauge theories has led to a remarkable result which is the description of the Higgs and the gauge fields in an equal footing. Among the many attempts to reformulate the classical standard model are those using the discrete space approach [1]-[11], where a gauge invariant Lagrangian for bosonic and fermionic sectors has been successfully constructed. However, the quantized version of the model in this scheme remains unclear. The goal of this paper is to reformulate a BRS quantized standard model in the context of a new superspace formalism where the Yang-Mills, Higgs and ghosts fields are unified in one super connection and the BRS transformations are obtained from horizontality like condition.

FORMALISM

In what follows we introduce a discrete superspace $M_4 \otimes Z_2 \otimes S^1$ in which M_4, Z_2 and S^1 stand for the Minkowski, discrete and Grassman spaces respectively. We define a generalized 1-super form A as follows:

$$\mathrm{A} = V_\mu dx^\mu + A_\theta d\theta + A_h \chi^h \tag{1}$$

where dx^μ, $d\theta$ and χ^h represent a super basis of the super discrete one form space such that the super valued functions V_μ, A_θ and A_h are given in six components vectors by:

$$V_\mu = \left(A_\mu, \theta\delta_{BRS}A_\mu, 0,0,0\right),\ A_h = \left(\varphi_h, 0, \theta\delta_{BRS}\varphi_h, 0,0,0\right),\ A_\theta = \left(0,0,0,\overline{C}, \theta\delta_{BRS}\overline{C}, 0\right) \tag{2}$$

and verify the following non vanishing scalar product relations:

$$\langle dx^\mu, dx^\nu \rangle = g^{\mu\nu},\ \langle \chi^g, \chi^h \rangle = \eta^{gh},\ \langle d\theta, d\theta \rangle = g^{\theta\theta} = 1 \tag{3}$$

CP1006, *High Energy Physics and Applications*, edited by J. Ellis, S. Nasri, and E. Malkawi

The metric η^{gh} is taken to be proportional to $\delta^{g,h^{-1}}$ and $g^{\mu\nu} = diag(1,-1,-1,-1)$, the δ_{BRS} stands for BRS transformations and θ is a Grassman variable. We introduce the following generalized exterior derivative dA such that:

$$dA = \Omega_{\nu\mu} dx^{\nu} \wedge dx^{\mu} + \Omega_{h\mu} \chi^{h} \wedge dx^{\mu} + \Omega_{\theta\mu} d\theta \wedge dx^{\mu} + \Omega_{\theta g} d\theta \wedge \chi^{g} + \Omega_{hg} \chi^{h} \wedge \chi^{g} \tag{4}$$

where

$$\Omega_{\nu\mu} = \frac{1}{2}\left(\partial_{\nu} V_{\mu} - \partial_{\mu} V_{\nu}\right), \Omega_{h\mu} = \hat{\partial}_{h} V_{\mu} - \partial_{\mu} A_{h}, \Omega_{\theta\mu} = \partial_{\theta} V_{\mu} - \partial_{\mu} A_{\theta}, \Omega_{\theta g} = \partial_{\theta} A_{g} - \hat{\partial}_{g} A_{\theta} \tag{5}$$

and

$$\Omega_{hg} = \hat{\partial}_{h} A_{g} - C^{p}_{h,g} R_{h \bullet g} A_{p} \tag{6}$$

("$\bullet$" is the internal operation of the discrete Z_2 group).We construct also a generalized exterior product $\overset{*}{\wedge}$ such that:

$$A \overset{*}{\wedge} A = \Lambda_{\mu\nu} dx^{\mu} \wedge dx^{\nu} + \Lambda_{\mu\theta} dx^{\mu} \wedge d\theta + \Lambda_{\mu g} dx^{\mu} \wedge \chi^{g} + \Lambda_{\theta g} d\theta \wedge \chi^{g} + \Lambda_{gh} \chi^{g} \wedge \chi^{h} \tag{7}$$

where

$$\Lambda_{\mu\nu} = \frac{1}{2}\left[V_{\mu}, V_{\nu}\right]_{*}, \ \Lambda_{\mu\theta} = V_{\mu} * A_{\theta}, \ \Lambda_{\mu g} = V_{\mu} * A_{g}, \ \Lambda_{\theta g} = A_{\theta} * A_{g}, \ \Lambda_{gh} = A_{g} * A_{h} \tag{8}$$

Therefore, a generalized 2-superform field strength F can be defined as:

$$F = dA + A \overset{*}{\wedge} A \tag{9}$$

Direct simplifications show that F can have the form:

$$F = F_{C} + F_{Q} \tag{10}$$

where F_C and F_Q stand for the classical and quantum field strength respectively. We define the quantum Lagrangian density (Gauge fixing and Fadeev-Popov terms) L_Q as:

$$L_{Q} = \int d\theta \ Tr\langle F_{Q}, F_{Q} \rangle_{\otimes} \tag{11}$$

Here ``Tr" refers to the trace over the Dirac matrices and the generalized scalar product $\langle \ , \ \rangle_{\otimes}$ is such that:

$$\langle F_{Q}, F_{Q} \rangle_{\otimes} = \langle F_{Q}, M\tilde{F}_{Q}^{+} \rangle \tag{12}$$

where

$$F_{Q} = F_{\mu\theta} dx^{\mu} \wedge d\theta + F_{\theta g} d\theta \wedge \chi^{g}, \quad \tilde{F}_{Q} = \tilde{F}_{\mu\theta} dx^{\mu} \wedge d\theta + \tilde{F}_{\theta h} d\theta \wedge \chi^{h} \tag{13}$$

with

$$\begin{gathered} F_{\mu\theta} = \partial_{\mu} A_{\theta} - \partial_{\theta} V_{\mu} + \left[V_{\mu}, A_{\theta}\right]_{*}, \ F_{\theta h} = \partial_{\theta} A_{h} - \hat{A}_{\theta} + A_{\theta} * A_{h} \\ \tilde{F}_{\mu\theta} = \partial_{\mu} \tilde{A}_{\theta} - \partial_{\theta} \tilde{V}_{\mu} + \left[\tilde{V}_{\mu}, \tilde{A}_{\theta}\right]_{*}, \ \tilde{F}_{\theta h} = \partial_{\theta} \tilde{A}_{h} - \hat{\tilde{A}}_{\theta} + \tilde{A}_{\theta} * \tilde{A}_{h} \end{gathered} \tag{14}$$

and

$$M = \begin{pmatrix} 0 & 1 & 1 & 0 & 0 & 0 \\ 0 & 0 & 0 & 0 & 1 & 0 \\ 0 & 0 & 0 & 0 & 1 & 0 \\ 0 & 0 & 0 & 0 & 0 & 0 \\ 0 & -1 & 0 & 0 & 0 & 0 \\ 0 & 0 & -1 & -1 & 0 & 0 \end{pmatrix} \tag{15}$$

The choice of the matrix M is not arbitrary, it allows the elimination of the unwanted terms. Here the star-* product is defined such that:

$$(a_1,a_2,a_3,a_4,a_5,a_6)*(b_1,b_2,b_3,b_4,b_5,b_6)=(a_1b_1,a_2b_2,a_3b_3,a_4b_4,a_5b_5,a_6b_6) \tag{16}$$

and

$$[V_\mu,V_\nu]_* = V_\mu * V_\nu - V_\nu * V_\mu \tag{17}$$

The new components $\widetilde{V}_\mu$, $\widetilde{A}_h$, $\widetilde{A}_\theta$ and $\widetilde{\hat{A}}_\theta$ have the following expressions:

$$\widetilde{V}_\mu = (\delta_{BRS}A_\mu, \theta A_\mu, 0,0,0,0), \ \widetilde{A}_h = (\delta_{BRS}\varphi_h, 0, \theta\varphi_h, 0,0,0), \ \widetilde{A}_\theta = (0,0,0,\delta_{BRS}\overline{C}, \theta\overline{C}, 0) \tag{18}$$

and

$$\widetilde{\hat{A}}_\theta = (0,0,0,\delta_{BRS}\overline{C},0,\theta\overline{C}) \tag{19}$$

In order to reproduce the quantized Lagrangian of the standard model, we have to make the following simplifications:

$$A_\mu(x,e)=T^a A_\mu^a(x), A_\mu(x,h)=0=0 \tag{20}$$

and

$$\varphi_h(x,e)=\varphi_h^+(x,h)=\frac{i}{\sqrt{2}}\sum_{a=0}^{3}T^a\varphi^a\begin{pmatrix}0\\1\end{pmatrix}, \varphi_e(x,e)=\varphi_e(x,h)=0 \tag{21}$$

where {e,h} (e is the neutral element, h≠e and $h=h^{-1}$)are the elements of the Z_2 discret group. It is worth to mention that the exterior derivatives of the 1-superforms basis are constrainted to verify the following equations:

$$ddx^\mu = 0, \ \ dd\theta = 0, \ \ d\chi^g = -C_{p,h}^g \chi^p \wedge \chi^h \tag{22}$$

The quantities $C_{p,h}^g$ are the pure discrete space structure constants. The 1-form discrete space basis χ^g has a non trivial commutativity relation namely:

$$\chi^g f(x,h) = (R_g f(x,h))\chi^g \tag{23}$$

where the action of the operator R_g is such that

$$R_g f(x,h) = f(x, g\bullet h), \qquad g,h\in Z_2 \tag{24}$$

Moreover, the super valued functions V_μ, A_h and A_θ are supposed to have the following properties:

$$(V_\mu)^* = -V_\mu, \ \ A_h^+ = R_g A_{h^{-1}}, \ \ (A_\theta)^* = A_\theta \tag{25}$$

(h^{-1} is the inverse of the element h). Now, imposing the nilpotency property of the generalized exterior product ($dd\mathrm{A}=0$), we end up with the following conditions:

$$C_{h,g}^p R_{h\bullet g}\hat{\partial}_p V_\mu = 0, \quad R_{h\bullet g} C_{h,g}^p \hat{\partial}_p A_\theta = 0 \tag{26}$$

Notice that to ensure the elimination of the unwanted terms in the final quantum Lagrangian, a generalized discrete derivative $\hat{\partial}_g$ is introduced and defined such that:

$$\hat{\partial}_g A_h = \partial_g A_h, \quad \hat{\partial}_g A_\mu = \partial_g A_\mu, \quad \hat{\partial}_g A_\theta = \partial_g \hat{A}_\theta \tag{27}$$

with ∂_g is the ordinary discrete space derivative such that

$$\partial_g f(x,h) = f(x,h) - R_g f(x,h) \tag{28}$$

and

$$\hat{A}_\theta = \left(0,0,0,\bar{C},0,\theta\delta_{BRS}\bar{C}\right) \tag{29}$$

Furthermore, we associate to each super valued functions V_μ, A_θ and A_h the following characters ξ_μ, ξ_θ and ξ_h respectively defined as:

$$\xi_\mu = \sum_{j=1}^{6}\left(V_\mu\right)_j, \quad \xi_\theta = \sum_{j=1}^{6}\left(\bar{A}_\theta\right)_j, \quad \xi_h = \sum_{j=1}^{6}\left(A_h\right)_j \tag{30}$$

where $$\bar{A}_\theta = \left(0,0,0,C,\theta\delta_{BRS}C,0\right) \tag{31}$$

Now, we construct the associated fields strength $\varpi_{\mu\theta}$, $\varpi_{g\theta}$ and $\varpi_{\theta\theta}$ such that:

$$\varpi_{\mu\theta} = \partial_\mu\xi_\theta - \partial_\theta\xi_\mu + \left[\xi_\mu,\xi_\theta\right], \quad \varpi_{g\theta} = \partial_g\xi_\theta - \partial_\theta\xi_g - \frac{i}{2}\left\{\xi_g,\xi_\theta\right\} \quad \varpi_{\theta\theta} = \partial_\theta\xi_\theta + \left\{\xi_\theta,\xi_\theta\right\} \tag{32}$$

in order to get the BRS transformations of the various generic fields, we impose the following horizontality condition:

$$\varpi_{\mu\theta}\Big|_{\theta=0} = \varpi_{g\theta}\Big|_{\theta=0} = \varpi_{\theta\theta}\Big|_{\theta=0} = 0 \tag{33}$$

which is very different from those of ref.[7].Direct simplifications of eqs.(31) lead to:

$$\delta_{BRS}W_\mu^\pm = \partial_\mu C^\pm \mp igW_\mu^\pm(C^z\cos\theta_w - C^A\sin\theta_w) \pm igC^\pm\left(Z_\mu\cos\theta_w - A_\mu\sin\theta_w\right)$$

$$\delta_{BRS}Z_\mu = \partial_\mu C^z + i\frac{g\cos\theta_w}{2}(W_\mu^+C^- - W_\mu^-C^+)$$

$$\delta_{BRS}A_\mu = \partial_\mu C^A - i\frac{g}{2}(W_\mu^+C^- - W_\mu^-C^+)\sin\theta_w$$

$$\delta_{BRS}\varphi^\pm = \frac{g}{2}\left[\pm i(C^z - C^A tg\theta_w)\varphi^\pm \pm \frac{i}{\sqrt{2}}C^\pm\left(v + H + i\varphi^z\right)\right]$$

$$\delta_{BRS}\varphi^z = \frac{g}{2}\left[\ \frac{1}{\sqrt{2}}(C^-\varphi^+ - C^+\varphi^-) - (C^A + C^z tg\theta_w)(v+H)\cos 2\theta_w\ \right]$$

$$\delta_{BRS}H = \frac{g\cos 2\theta_w}{2}(C^A + C^z tg\theta_w)\varphi^z \tag{34}$$

with

$$W_\mu^\pm = \frac{1}{\sqrt{2}}\left(A_\mu^1 \pm iA_\mu^2\right) \quad A_\mu. = A_\mu^0\cos\theta_w - A_\mu^3\sin\theta_w, \quad Z_\mu. = A_\mu^0\sin\theta_w + A_\mu^3\cos\theta_w \tag{35}$$

and $$\varphi^z = \varphi^3, \quad \varphi^0 = -i(v+H), \quad \varphi^\pm = (\varphi^1 \pm \varphi^2)/\sqrt{2} \tag{36}$$

'v'is the vacuum expectation value. After straightforward simplifications, the quantized Lagrangian density L_Q takes the form:

$$L_Q = L_{GF} + L_{FP} \tag{37}$$

where

$$L_{GF} = -\xi'\left(\partial^\mu A_\mu\right)^2 - \xi\left|\partial^\mu W_\mu^+ - i\frac{M_w}{\xi}\varphi^+\right|^2 - \frac{\eta}{2}\left(\partial^\mu Z_\mu - \frac{M_z}{\eta}\varphi_z\right)^2 \tag{38}$$

and

$$
\begin{aligned}
L_{FP} = {} & i\xi' \overline{C}^A \partial^\mu \left[\partial_\mu C^A - \frac{ie}{2}\left(W_\mu^+ C^- - W_\mu^- C^+\right)\right] + i\eta \overline{C}^z \left\{ \frac{ie\cos\theta_w}{2} \partial^\mu \left(W_\mu^+ C^- - W_\mu^- C^+\right) + \partial^\mu \partial_\mu C^A \right. \\
& \left. - i\frac{gM_z}{2\sqrt{2}\eta}\left[\left(C^-\varphi^+ - C^+\varphi^-\right) + \sqrt{2}\left(C^A + C^z tg\theta_w\right)\left(v+H\right)\cos 2\theta_w\right]\right\} + \left\{ i\xi \overline{C}^- \left\{ \partial^\mu \left[-ieW_\mu^+ \left(C^z \cot g\theta_w - C^A\right)\right.\right.\right. \\
& \left.\left.\left. + \partial^\mu \partial_\mu C^+ + ieC^+\left(Z_\mu \cot g\theta_w - A_\mu\right)\right] + \frac{gM_w}{2\sqrt{2}\xi}\left[C^+\left(v+H+i\varphi_z\right) + \sqrt{2}\varphi^+\left(C^A + C^z tg\theta_w\right)\right]\right\} + c.c \right\}
\end{aligned}
\tag{39}
$$

(c.c stands for complex conjugate)

CONCLUSION

Fadeev-Popov and gauge fixing terms of the quantized version of the standard electroweak model are obtained by constructing an extended discrete superspace formalism with a new mathematical approach different from those proposed in ref.[7]. Moreover, a generalized scalar product is defined and the BRS transformations of the various dynamical fields are derived from an horizontality like condition.

ACKNOWLEDGMENTS

We are very grateful to the organizers of the UAE-CERN Worhshop for their kind hospitality which we have during our stay at El Ain and the Algerian Ministry of education and research for financial support.

REFERENCES

1. K.Morita and Y.Okumura, *Prog.Theor.Phys* **95**, 227-248 (1996).
2. Y.Okumura, *Eur.Phys. J.C.***1**, 711-715 (1998).
3. A.Sitarz, *J.Geom.Phys.* **15**, 123-136 (1995).
4. A.Sitarz, *Phys.Lett.B***8** , 311-314 (1993).
5. A.Dimakis and F.M.Hoissen, *J.Math.Phys.* **36**, 3771-3791 (1995).
6. Y.Okumura, *Phys.Rev.D***54**, 4114-4120 (1996).
7. Y.Okumura, *J.Math.Phys* **41**, 1788-1800 (2000).
8. H.Kase, K.Morita and Y.Okumura, *Int.Jour.Mod.Phys.A* **16**, 3203-3216 (2001).
9. H.Kase, K.Morita, Y.Okumura and U.Eizou, *Prog.Theo. Phys.***109**, 663-685 (2003).
10. V.Nazaryan and C. E. Carlson, *Phys.Rev. D***71**, 025019 (2005).
11. K.Morita and Y.Okumura, *.Prog.Theo. Phys.***91**, 975-990 (1994).

Eta Structure Function via Photo-production

N. Ghahramany, M. Ghanaatian and A. Bazrafshan

Dept. of Physics, Shiraz University, Shiraz 71454, Iran

Abstract. By using isobar model and the latest data, the structure function of Eta particle is calculated in the process of photo-production. This calculation is carried out for photon energy range of 0.750-3 GeV. The variations of the structure function are given in terms of scattering angle for different incoming photon energy.

Keywords: Eta meson, Cross-section, Photo-production, Electro-production, Structure function
PACS: 14.40.Aq, 13.85.Lg, 13.40.Gp

INTRODUCTION

In the late sixties many effort was made to investigate the eta photo-production process, however they are so limited due to large error bars in their data [1-4]. In late eighties some measurements were made again with relative improvement but not enough in the reduction of uncertainties to be used in reliable calculations [5, 6].

The most recent data collected in 2003 at experiment BONN-ELSA-CB Detector/Collaboration carry acceptable error bars. Therefore, these experimental data are used here to investigate the total cross section and the structure function of eta particle in the interaction $\gamma\, p \rightarrow \eta\, p$. The variations of the eta structure functions are obtained in terms of the incoming photon energy in the range of 0.75-3.00 (GeV). The angular distribution of eta structure function is also studied.

KINEMATICS OF ETA PHOTO-PRODUCTION

The electromagnetic production of eta is possible via two different interactions, photo- and electro-production, namely:

$$\gamma(q) + N(p_i) \rightarrow \eta(k) + N(p_f) \qquad (1)$$

$$e(e) + N(p_i) \rightarrow e'(e') + \eta(k) + N(p_f) \qquad (2)$$

where the four-momentum in the parentheses are defined as: $e = (E, \vec{p}_e)$ for the incident electron, $e' = (E', \vec{p}'_e)$ for the scattered electron, $p_i = (E_i, \vec{p}_i)$ for the target nucleon, $p_f = (E_f, \vec{p}_f)$ for the recoiled nucleon, $k = (\omega, \vec{k})$ for the produced eta and the four-momentum of the virtual photon exchanged in electro-production is given by $q = (\upsilon, \vec{q}) = e - e'$ and for photo-production $q^2 = 0$.

In the c.m. system for $\gamma\, N \rightarrow \eta\, N$ interaction, the initial and final momenta of the involved particles can be expressed in terms of the total center of mass energy $w = \sqrt{s}$ and momentum transfer squared q^2 as follow:

CP1006, *High Energy Physics and Applications*, edited by J. Ellis, S. Nasri, and E. Malkawi

$$|\vec{q}| = |\vec{p}_i| = \frac{1}{2w}\sqrt{[(w+M_i)^2+Q^2][(w-M_i)^2+Q^2]} \tag{3}$$

$$|\vec{k}| = |\vec{p}_f| = \frac{1}{2w}\sqrt{[(w+M_f)^2-m^2][(w-M_f)^2-m^2]} \tag{4}$$

Where $Q^2 = -q^2 = \vec{q}^2 - \upsilon^2$ and M_i and M_f are initial and final mass of nucleon and m is the mass of eta.

In photo-production, the relation between Lab. photon energy E_γ^{lab} and the total center of mass energy w is given as:

$$E_\gamma^{lab} = \frac{w^2-M_i^2}{2M_i} \tag{5}$$

The five-fold differential cross-section for electro-production is expressed as [7-10]:

$$\frac{d^5\sigma}{d\Omega' dE' d\Omega_\eta} = \Gamma\frac{d^2\sigma}{d\Omega_\eta^*} \tag{6}$$

With the flux of the virtual photon field is given by:

$$\Gamma = \frac{\alpha}{2\pi^2}\frac{E}{E'}\frac{K}{Q^2}\frac{1}{1-\varepsilon} \tag{7}$$

Where $K = E_\gamma^{lab}$ is the photon equivalent energy and $\varepsilon = \left(1 + 2\frac{\vec{q}^2}{Q^2}\tan^2(\frac{\theta_e}{2})\right)^{-1}$ stand for transverse polarization parameter of the virtual photon.

For an un-polarized target without recoil polarization, the virtual photon differential cross-section is:

$$\frac{d^2\sigma}{d\Omega_\eta^*} = \frac{d^2\sigma_T}{d\Omega_\eta^*} + \varepsilon\frac{d^2\sigma_L}{d\Omega_\eta^*} + \sqrt{2\varepsilon(1+\varepsilon)}\frac{d^2\sigma_{LT}}{d\Omega_\eta^*}\cos(\varphi) + \varepsilon\frac{d^2\sigma_{TT}}{d\Omega_\eta^*}\cos(2\varphi) \tag{8}$$

In contrast to photo-production in which the energy and momentum of the photon is given by $q^2 = \upsilon^2 - \vec{q}^2 = 0$, in the case of electro-production, the virtual photon make it possible for the real photon to have an independent energy and momentum variable to provide information about the spatial hadronic structure function.

In photo-production experiment only two kinematics variables are enough to express the kinematics situation of the system. These two variables are usually chosen to be the scattering angle θ and the Lab. energy of photon E_γ^{lab}. Therefore in photo-production process the longitudinal components of the cross-section vanish and the obtained structure function correspond to the transverse component of the cross-section which is determined by considering the polarization observables.

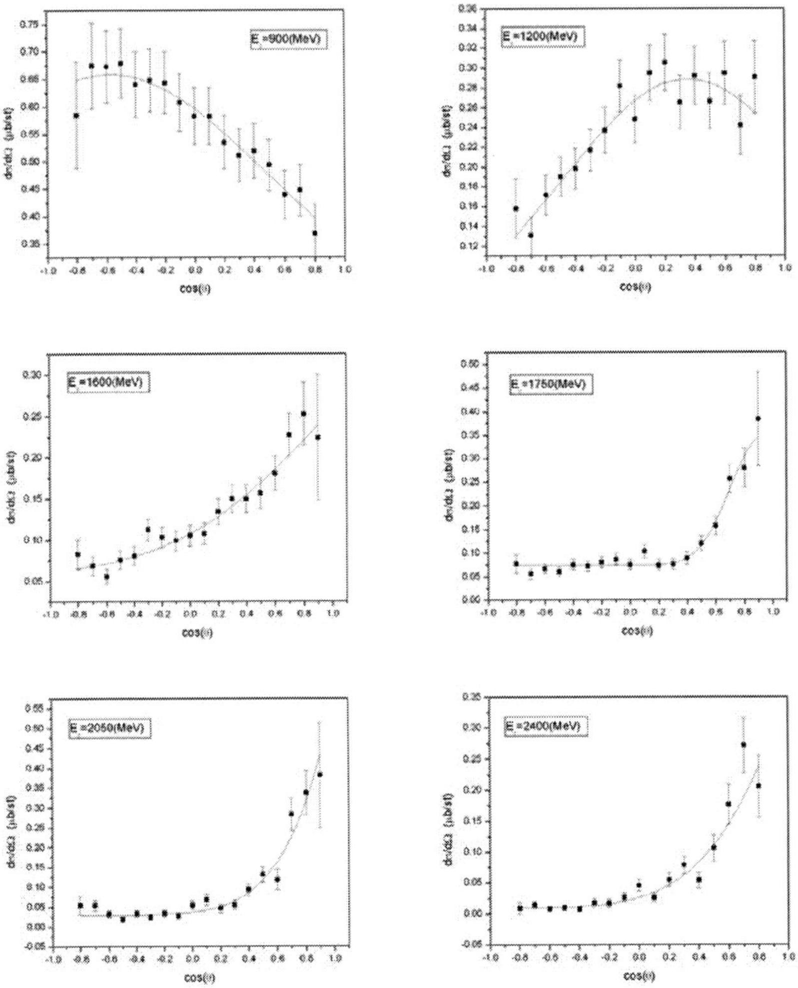

FIGURE 1. Sample graphs of the angular distribution of the differential cross-section of eta photo-production at different known photon energy E_γ^{lab}.

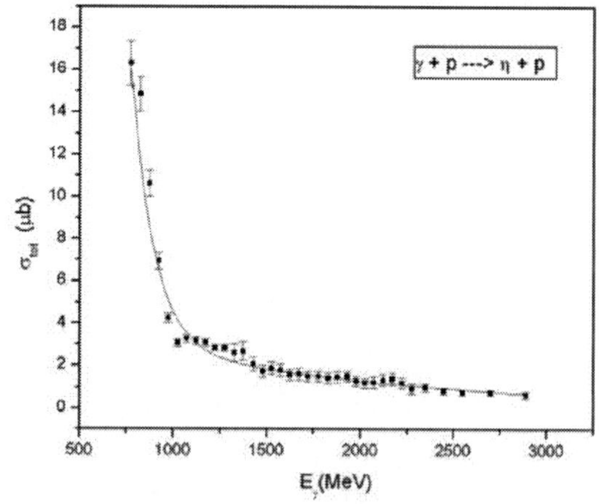

FIGURE 2. Eta photo-production total cross-section as a function of E_{γ}^{lab}.

CONCLUSION

The investigation of the electro-production and photo-production of eta meson is increasingly becoming interesting especially in medium energy physics. Since eta is an isospin singlet more complete information about the nucleon structure could be obtained compared to pion production. Therefore doing such experiments is in the list of future experimental performances of many laboratories. In this research by using the latest available data the variation of the total cross-section is obtained in terms of $E_{\gamma}^{lab}$and $\cos(\theta)$ (Fig. 1 and Fig. 2) and are in good agreement with the previous performed investigations.

ACKNOWLEDGMENTS

Partial support of the research council of Shiraz University is appreciated.

REFERENCES

1. B. Delcourt et al., *Phys. Lett.* **B 29**, 75 (1967).
2. C. Bacci et al., *Phys. Rev. Lett.* **20**, 571 (1968).
3. E. D. Bloom et al., *Phys. Rev. Lett.* **21**, 1100 (1968).
4. C. A. Heusch et al., *Phys. Rev. Lett.* **25**, 1381 (1970).
5. S. Dytman et al., *Bull. Amer. Phys. Soc.* **35**, 1679 (1990).
6. S. Homma et al., *J. Phys. Jpn.* **57**, 828 (1988).
7. N. Ghahramany, K. Rostami and M. Ghanaatian, *Communic. in Theoretical Phys.*, **Vol. 41 No. 2** (2004).
8. N. Ghahramany and M. Ghanaatian, *Iranian Journal of Science and Technology*, **Transaction** A **Vol. 29 No. A2** (2005).
9. N. Ghahramany, M. Ghanaatian and P. Dehghanipoor, *Iranian Physical Journal*, **Vol. 1** (2007).
10. D. Drechsel and L. Tiator, *J. Phys.* **G 18**, 449 (1992).

Debye Layer Confinement of Nucleons in Nuclei for Heavy Nuclear Synthesis

N. Ghahramany, M. Ghanaatian

Dept. of Physics, Shiraz University, Shiraz 71454, Iran

Abstract. Following the laser ablation studies leading to a theory of nuclei confinement by a Debye layer mechanism, we present here numerical evaluations for the known stable nuclei where the Coulomb repulsion is included as a rather minor component especially for lager nuclei. It is noticed that the well known empirical nuclear density of $2 \times 10^{38} cm^{-3}$ follows for a nucleon number A< 60 for iron. The crucial change of the Fermi energy into the relativistic branch for the nucleons results for a density at $3 \times 10^{39} cm^{-3}$ which density corresponds to the Debye layer equilibrium near uranium above which the Fermi energy will not permit any nucleation. It is speculated whether the range between both densities at the big bang expansion at temperatures of few 100 keV and at 200 seconds after the big bang is resulting in a nuclear-chemical Boltzmann equilibrium for the generation of the endothermic nuclear generation. The question is then open, what additional nuclear expansion forces are acting in the elements between uranium and iron during this equilibrium reactions in nuclei such that the empirical nuclear density is produced.

Keywords: Endothermic nuclear synthesis, Fermi energy, Relativistic change, Surface tension, Debye length.
PACS: 26.35.+c, 98.80.Ft, 25.75.Nq, 12.38.Mh

INTRODUCTION

For many years astrophysicists have been wondering how endothermic synthesis of nuclei heavier than Iron could be produced resulting in the existing abundance in our Universe. The exothermic nucleation process of elements, lighter than Iron, could take place via fusion reactions [1]. During the last 15 years research papers have been published addressing this question with new approach [2-6] following the theory of nonlinear forces and double layers in laser produced plasmas [7-10]. The first step from these classical plasma conditions to the degenerate electrons in a metal by changing temperature into Fermi energy for the Debye length resulted in surface energy and the work function of metals. This quantum theory of surface tension led to values agreeing with measurements [11] differing from the usual explanation of surface tension [12]

Following these results, this was generalized to the case of protons and neutrons instead of electrons with their Fermi energy instead of the temperature in the Debye layer. A first convincing semi-empirical result is when taking the averaged density of nucleons in nuclei of $n = 2 \times 10^{38} cm^{-3}$, one arrives at a Debye length of the value of the measured [6] 3 *fm* decay of the nuclear density at their surface as measured by Hahn et al [13] and Hofstadter et al [14]. When dropping the empirical value of n and calculating the surface energy form the Debye layer, it was noted [2-5] that the compression of the nucleons is about compensated by the surface energy just about at the density of the nuclei. In this case the compression energy was just only that of the

CP1006, *High Energy Physics and Applications*, edited by J. Ellis, S. Nasri, and E. Malkawi

Fermi energy of the nucleons. We present here the inclusion of the Coulomb repulsion energy of the nuclei and take into account some updating of the preceding formulations. Since the number of proton charges has to be taken individually, we have to follow up the energy values for each nucleus as given in the following separately for the subrelativistic and for the relativistic range of the Fermi energy. The calculations are done in such a way that the compensation of the interior energy of the unexcited nucleus (enthalpy) with the surface energy is the starting point and the resulting density n of the nucleus is calculated.

THE SURFACE TENSION AND ENERGY IN A DEBYE LAYER OF NUCLEI

The thickness of the double layer formed in the homogeneous plasma expanding into vacuum [3,8] is given by the Debye length [12]:

$$\lambda_D = [KT/(4\pi e^2 n_e)]^{\frac{1}{2}} \tag{1}$$

Where K is the Boltzman constant, T is the plasma electron temperature, e and n_e are electron charge and density of plasma respectively. The surface tension is defined as the electric field energy within the volume V (given by layer surface, S times Debye length λ_D) per unit area of layer [8]:

$$\sigma_e = 0.27T^2/(8\pi e^2 \lambda_D) \tag{2}$$

The surface tension for nuclei is given by replacing temperature by Fermi energy of protons and neutrons [5]. Therefore we have

$$\sigma_n = 0.27E_F^2/(8\pi e^2 \lambda_D) \tag{3}$$

The Fermi energy is generally given [9]

$$E_{F=}\frac{(3/\pi)^{2/3}}{4}\frac{h^2 n^{2/3}}{2m}\left\{\left(\frac{\lambda_c}{2}\right)[n + 1/(\lambda_c/2)]^{1/3}\right\}^{-1} \tag{4}$$

where n is the nucleon density and m is the nucleon mass and λ_C is the Compton wavelength (h/mc) for nucleons. This energy splits into two branches, relativistic and subrelativistic,

$$E_{F,sub} = \frac{\left(\frac{3}{\pi}\right)^{\frac{2}{3}}}{4}\frac{h^2 n^{\frac{2}{3}}}{2m} \quad (5)$$

$$E_{F,rel} = \frac{\left(\frac{3}{\pi}\right)^{\frac{2}{3}}}{4} hcn^{1/3} \quad (6)$$

The surface energy of the nucleus [5] is give

$$E_{surf} = 0.27\frac{[3A(4\pi)^{1/2}]3^{1/2}E_F^{3/2}}{\pi^{1/2}2^{5/2}n^{1/6}e} \quad (7)$$

In order to compare the surface energy and the Fermi energy of the nucleus, their ratio is considered, giving,

$$\gamma = \frac{E_{surf,sub}}{AE_{F,sub}} = 0.27\frac{3^{3/2}hn^{1/6}}{2^{10/3}\pi^{1/2}m^{1/2}A^{1/3}e} \quad (8)$$

$$\gamma' = \frac{E_{surf,rel}}{AE_{F,rel}} = 0.27\frac{3^{3/2}}{2^{7/3}\alpha^{1/2}A^{1/3}} \quad (9)$$

where $\alpha = \frac{2\pi e^2}{hc}$ is the fine structure constant.

From equation (8) we conclude that if the density is too low, then the nucleus confinement by the Debye layer is not possible. The nucleus will be stable only if the ratio in (8) is equal to one. At relativistic densities above that of sub-relativistic case, we have

$$\frac{E_{surf,rel}}{AE_{F,rel}} = \frac{6.514}{A^{1/3}} \quad (10)$$

which does not depend upon mass and density of the nucleon.

Debye layer was approximately compensating the inner energy of nuclei for a confinement of the nucleons. Values for $A \leq 60$ for the stable nuclei are calculated using γ-values equal unity, to result in the nuclear density and the corresponding value of $A > 60$ for the stable nuclei.

It is interesting that the resulting nuclei from hydrogen until iron arrive at lower nuclear density than the well known, $n_0 = 2 \times 10^{38} cm^{-3}$, approximate value of all nuclei. This fact indicates that either the Debye layer model is not applicable for explaining the nucleon confinement for the nuclei lighter than iron, or there are further repulsive forces acting in these nuclei.

The further point of interest is that at the heavy nuclei up to curium, the subrelativistic γ-value results in the same value γ' as the relativistic branch. This is important as the nuclear density of $n^* = 3 \times 10^{39} cm^{-3}$ is just the value where the subrelativistic Fermi energy changes into the relativistic branch. At higher nuclear densities, there is no nucleation possible at thermal equilibrium and only a homogeneous soup of particles is possible as a quark-gluon plasma or as assumed in a neutron star with such high density as clarified before [2-6].

CONCLUSION

We assume that when matter with this homogeneous structure is expanding as in a big bang or at the expansion of a neutron star in a supernova, that the density range between n_0 and n^* - a 15 times variation of the nuclear density - is the range where the nuclear-chemical equilibrium is due to an up and down of nuclear reactions where the Boltzmann factor determines the final distribution for the Z-dependence of the nuclei above $A=60$. This distribution was agreeing with the measured abundance in the universe and agreed to an explanation of the magic numbers of nuclei by a 3^n-series ($n=1,2,3,\ldots$) reflecting a threefold quark structure of stable shells in nuclei [4].

The action of additional forces in nuclei during the 15 times density variation or for the light nuclei below iron is not followed up here and a point of further speculation and research.

ACKNOWLEDGMENTS

Partial support of the research council of Shiraz University is appreciated.

REFERENCES

1. R. Kippenhahn and A. Weigert, *Stellar Structure and Evolution*, Springer, Heidelberg, 1990.
2. H. Hora, *Plasma Model for Surface Tension of Nuclei and the Phase Transition to the Quark Plasma*, Report CERN-PS/DL-Note-91/05, August (1991)
3. H. Hora, et al, *High Power Laser Ablation*, Proceeding of SPIE, **Vol. S448**, 1190-1200, SPIE, Bellingham, WA, 2004.
4. H. Hora, G.H. Miley and F. Osman, *Astrophysics and Space Science* **298**, 247-253, (2005).
5. F. Osman, N. Ghahramany, and H. Hora, *Laser and Particle Beams* **23**, 461-466, (2005).
6. H. Hora, *Laser and Particle Beams* **24**, 35-40 (2006).
7. H. Hora, *Phys. Fluids* **12**, 181 (1969).
8. H. Hora, *Plasmas at High Temperature and Density*, Springer, Heidelberg, 1991; H. Hora, *Laser Plasma and the Nonlinearity Principle*, SPIE Press, Bellingham WA, 2000.
9. S. Eliezer, A. J. Ghatak, H. Hora, and E. Teller, *Fundamentals of Equation of State*, World Scientific, Singapore, 2002.
10. H. Hora, P. Lalousis and S. Eliezer, *Phys. Rev. Letters* **53**,1650 (1984); S. Eliezer and H. Hora, *Physics Reports* **172**, 339 (1989).
11. H. Hora et al, *On Surface Tension in Plasmas*, IEEE Trans. Plasma Sc. PS-17, 284-289, 1989.
12. N. G. Laud and W. J. Kohn, *Phys. Rev.* **B1**, 4555(1970).
13. B. Hahn, D.G. Ravenhall and R. Hofstadter, *Phys. Rev.* **101**, 1131-1142 (1956)

Strong coupling constant and flavor thresholds in QCD

Ali N. Khorramian, S. Atashbar Tehrani, S. Taheri Monfared and H. Khanpour

Physics Department, Semnan University, Semnan, Iran

Abstract. NLO QCD analysis of the world unpolarized and polarized deep-inelastic experimental data were performed in order to extract the Λ_{QCD} and $\alpha_s(M_Z^2)$. In this paper we use the Λ_{QCD} which can be obtain from the polarized deeply inelastic scattering. By having the Λ_{QCD}, one can determine the $\alpha_s(M_Z^2)$. We compare all of different NLO measurements of the strong coupling constant in unpolarized and polarized deeply inelastic scattering.

Keywords: Strong coupling constant, DIS data
PACS: 12.38.Bx, 14.65.-q

INTRODUCTION

The coupling constant of the strong interactions, α_s, is one of the most fundamental parameters of nature which must be determined by experiment. Many new results, experimental studies as well as improved theoretical calculations, continue to be provided each year, such that regular updates of summaries of measurements of α_s are mandatory. Various QCD analysis of the world unpolarized and polarized deep-inelastic experimental data were performed in order to extract the Λ_{QCD} and $\alpha_s(M_Z^2)$. The world average for $\alpha_s(M_Z^2)$ which has been determined recently is $\alpha_s(M_Z^2) = 0.1189 \pm 0.0010$. In Ref. [1] a QCD analysis is performed for polarized structure function based on the Jacobi polynomials. In this paper we will use the value of Λ_{QCD} which extracted from the polarized deep-inelastic scattering data.

The plan of the paper is to give a brief expressions asymptotic freedom in QCD. Quark masses and thresholds, and also comparison of different NLO measurements of the strong coupling constant in unpolarized and polarized deeply inelastic scattering are summarized in the last section.

ASYMPTOTIC FREEDOM IN QCD

While QCD does not predict the actual size of α_s at a particular energy scale, its energy dependence is precisely determined. If the renormalized coupling $\alpha_s(\mu^2)$ can be fixed at a given scale μ^2, QCD definitely predicts the size of α_s at any other energy scale Q^2 through the renormalization group equation (RGE) [2]

$$Q^2 \frac{\partial \alpha_s(Q^2)}{\partial Q^2} = \beta(\alpha_s(Q^2)) \, . \tag{1}$$

CP1006, *High Energy Physics and Applications*, edited by J. Ellis, S. Nasri, and E. Malkawi

The perturbative expansion of the β function, including higher order loop corrections to the bare vertices of the theory, is calculated to complete 4-loop approximation [3]

$$\beta(\alpha_s(Q^2)) = -\beta_0\alpha_s^2(Q^2) - \beta_1\alpha_s^3(Q^2) - \beta_2\alpha_s^4(Q^2) - \beta_3\alpha_s^5(Q^2) - O(\alpha_s^6)\,. \tag{2}$$

β_0 can be obtained from a 1-loop calculation of any other physical quantity which depends on Q in perturbation theory. At leading order (LO), it is

$$\beta_0 = (11 - 2N_f/3)/4\pi = \frac{33 - 2N_f}{12\pi}\,, \tag{3}$$

where N_f is the number of flavors of quarks and N_c is the number of colors. The positive contribution, 11, comes mainly from the non-abelian diagrams in a quark-quark scattering process. The negative contribution, $-2N_f/3$, which weakens asymptotic freedom, comes from the fermion loop diagram. In these terms, the solution to the lowest order approximation to Eq. (1) can be written in terms of a single constant Λ which characterizes its strength at any given scale,

$$\alpha_s(Q^2) = \frac{1}{\beta_0 L}\,, \tag{4}$$

here $L = \ln(\frac{Q^2}{\Lambda^2})$. The $\Lambda(=\Lambda_{QCD})$ scale is called asymptotic scale parameter in QCD and depends on three factors: renormalization scheme, the number of active flavors, N_f and the order of approximation, e.g. LO, next-to-leading order (NLO), next-to-next-to-leading order (NNLO). The amount of Λ is not predictable from theory. It can be determined from QCD analysis of the hadron structure function [4].

The NLO solution for $\alpha_s(Q^2)$ is obtained by using the first two terms in the β function. There is some freedom of choice in its precise definition at NLO. One conventionally writes $\alpha_s(Q^2)$ in an expansion in powers of $1/L$, where the coefficient of $[1/L]^n$ is a polynomial in $ln(L)$. Keeping β_0 and β_1 allows us to determine the coefficients of $[1/L]^2$,

$$\alpha_s(Q^2) = \frac{1}{\beta_0 L} - \frac{b_1 \ln L}{(\beta_0 L)^2}\,, \tag{5}$$

where $b_1 \equiv \frac{\beta_1}{\beta_0}$ and $\beta_1 = \frac{153-19N_f}{24\pi^2}$.

The strong coupling up to NNLO is as following

$$\alpha_s(Q^2) = \frac{1}{\beta_0 L} - \frac{b_1 \ln L}{(\beta_0 L)^2} + \frac{1}{(\beta_0 L)^3}[b_1^2(\ln^2 L - \ln L - 1) + b_2]\,, \tag{6}$$

here $b_2 \equiv \frac{\beta_2}{\beta_0}$ and $\beta_2 = \frac{77139-15099N_f+325N_f^2}{3456\pi^3}$.

Since the strong coupling enters into the calculation of all processes beyond the leading order, it can be measured, in principle, in all processes involving hadronic particles. The universality of the function $\alpha_s(Q^2)$ provides the most powerful and decisive test of the validity of QCD. The universality of the measurements of running coupling at different scales can be tested by converting all the values to the same scale using the Eq. (6). The standard scale is usually chosen to be $Q^2 = M_Z^2$.

TABLE 1. Comparison of $\alpha_s(M_Z^2)$ values from NLO determined from unpolarized and polarized deep-inelastic scattering analysis.

NLO	$\alpha_s(M_Z^2)$	expt	theory	model	Ref.
CTEQ6	0.1165	± 0.0065			[6]
MRST03	0.1165	± 0.0020	± 0.0030		[7]
A02	0.1171	± 0.0015	± 0.0033		[8]
ZEUS	0.1166	± 0.0049		± 0.0018	[9]
H1	0.1150	± 0.0017	± 0.0050	$^{+0.0009}_{-0.0005}$	[10]
BCDMS	0.110	± 0.006			[11]
GRS	0.112				[12]
BBG	0.1148	± 0.0019			[13]
BB (pol)	0.113	± 0.004			[14]
SMC (pol)	0.121	± 0.002			[15]
ABFR (pol)	0.12	± 0.0098			[16]
AK(pol)	0.1141	± 0.0036			[1]

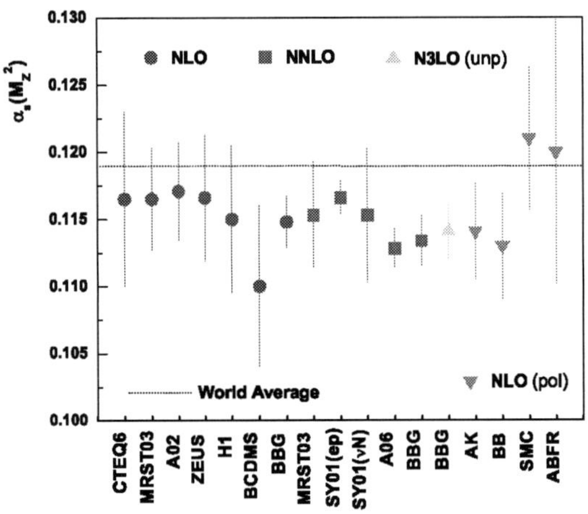

FIGURE 1. Comparison of different NLO, NNLO measurements of the strong coupling constant in unpolarized and polarized deeply scattering.

QUARK MASSES AND THRESHOLDS

Assuming that both the physical and renormalization scales Q^2 and μ^2 are larger than any other relevant energy or mass scale involved in the problem, the finite quark masses were neglected. This is, however, not entirely correct, since there are several QCD studies and α_s determinations at energy scales around the charm- and bottom-quark masses of about 1.5 and 4.7 GeV, respectively. In leading and next to leading order, the matching condition is $\alpha_s^{(N_f-1)} = \alpha_s^{(N_f)}$ [5]. In higher orders and the $\overline{MS}$ scheme, however, nontrivial matching conditions apply. Formally these are, if the energy evolution of α_s is performed in n^{th} order, of order $n-1$. According to Ref.[5, 17] the matching condition reads,

$$\frac{a'}{a} = 1 + C_2 a^2 + C_3 a^3 \,, \tag{7}$$

here $C_2 = -0.291667$, $C_3 = -5.32389 + (N_f - 1)0.26247$ and $a = \alpha_s^{(N_f)}/\pi$, $a' = \alpha_s^{(N_f-1)}/\pi$.

In table 1, we compare our NLO results for $\alpha_s(M_Z^2)$ and for unpolarized and polarized analysis. The reported value of Λ in [1] is 245 ± 58 MeV for polarized case. In figure 1, we compare the results of different NLO and NNLO QCD unpolarized and polarized DIS analysis for $\alpha_s(M_Z^2)$ with reported results from [1] and the world average for $\alpha_s(M_Z^2)$.

REFERENCES

1. S. Atashbar Tehrani and A. N. Khorramian, JHEP **0707** (2007) 048 [arXiv:0705.2647 [hep-ph]].
2. W. K. Tung, Hand Book of QCD, At the frontier of particle physics, vol. 2.
3. T. van Ritbergen, J. A. M. Vermaseren and S. A. Larin, Phys. Lett. B **400** (1997) 379 [arXiv:hep-ph/9701390].
4. J. Bluemlein, [arXiv:hep-ph/0706.2430].
5. K. G. Chetyrkin, B. A. Kniehl and M. Steinhauser, Phys. Rev. Lett. **79** (1997) 2184 [arXiv:hep-ph/9706430].
6. J. Pumplin, D. R. Stump, J. Huston, H. L. Lai, P. Nadolsky and W. K. Tung, JHEP **0207** (2002) 012 [arXiv:hep-ph/0201195].
7. A. D. Martin, R. G. Roberts, W. J. Stirling and R. S. Thorne, arXiv:hep-ph/0307262.
8. S. Alekhin, Phys. Rev. D **68** (2003) 014002 [arXiv:hep-ph/0211096].
9. S. Chekanov *et al.* [ZEUS Collaboration], Phys. Rev. D **67** (2003) 012007 [arXiv:hep-ex/0208023].
10. C. Adloff *et al.* [H1 Collaboration], Eur. Phys. J. C **21** (2001) 33 [arXiv:hep-ex/0012053];
 C. Adloff *et al.* [H1 Collaboration], Eur. Phys. J. C **30** (2003) 1 [arXiv:hep-ex/0304003].
11. A.C. Benvenuti *et al.* [BCDMS Collaboration], Phys. Lett. B **237** (1990) 592;
 A.C. Benvenuti *et al.* [BCDMS Collaboration], Phys. Lett. **B223** (1989) 485; Phys. Lett. **B237** (1990) 592.
 A.C. Benvenuti *et al.* [BCDMS Collaboration], Phys. Lett. B **237** (1990) 599.
12. M. Glück, E. Reya and C. Schuck, arXiv:hep-ph/0604116.
13. J. Blümlein, H. Böttcher and A. Guffanti, Nucl. Phys. B **774**, 182 (2007), arXiv:hep-ph/0607200.
14. J. Bluemlein and H. Bottcher, Nucl. Phys. B **636** (2002) 225 [arXiv:hep-ph/0203155].
15. B. Adeva *et al.* [Spin Muon Collaboration], Phys. Rev. D **58** (1998) 112002.
16. G. Altarelli, R. D. Ball, S. Forte and G. Ridolfi, Acta Phys. Polon. B **29** (1998) 1145 [arXiv:hep-ph/9803237].
17. S. Bethke, J. Phys. G **26** (2000) R27 [arXiv:hep-ex/0004021].

Polarized nucleon structure function up to NLO

S. Atashbar Tehrani and Ali N. Khorramian

Physics Department, Semnan University, Semnan, Iran
and
School of Physics, IPM (Institute for Studies in Theoretical Physics and Mathematics),
P.O.Box: 19395-5531, Tehran, Iran

Abstract. We present the results of our QCD analysis for helicity quark distribution and proton structure function $xg_1^p(x,Q^2)$. We use recently experimental data to parameterize our model. The analysis is based on the Jacobi polynomials expansion of the polarized structure functions. By having polarized proton structure functions based on the Jacobi polynomials method, determination of polarized deuteron structure function is possible. Our calculations for polarized deuteron structure function are in good agreement with the available experimental data.

Keywords: NLO Computation, Structure Function
PACS: 12.38.Bx, 14.65.-q

INTRODUCTION

The theoretical and experimental status on the spin structure of the nucleon has been discussed in great detail in several recent reviews (see, e.g., Refs. [1, 2, 3, 4]). During the recent years several comprehensive analysis of the polarized deep inelastic scattering (DIS) data, based on next-to-leading-order quantum chromodynamics, have presented. In these analysis the polarized parton density functions (PPDFs) are either written in terms of the well-known parameterizations of the unpolarized parton density functions (PDFs) or parameterized independently, and the unknown parameters are determined by fitting the polarized DIS data. Determination of parton distributions in a nucleon in the framework of quantum chromodynamics (QCD) always involves some model-dependent procedure. In the valon model, the proton consists of two "up" and one "down" valons. These valons thus, carry the quantum numbers of the respective valence quarks. In Ref. [5] the polarized valon model is applied to determine the quark helicity distributions and polarized proton structure function in the NLO approximation by using the Bernstein polynomial approach.

The extraction of the quark helicity distributions is one of the main tasks of the semi-inclusive deep inelastic scattering (SIDIS) experiments (HERMES [6], COMPASS [7], SMC[8]) with the polarized beam and target. Since recently experimental data are available from the HERMES collaboration [9] for the spin structure function g_1, therefore there is enough motivation to study and utilize the spin structure and quark helicity distributions extracted via the phenomenological model.

The plan of the paper is to give the theoretical background of the QCD analysis in two loops. The procedure of the QCD fit of the experimental data are written down in the last section.

CP1006, *High Energy Physics and Applications*, edited by J. Ellis, S. Nasri, and E. Malkawi

THE THEORETICAL BACKGROUND OF THE QCD ANALYSIS

Let us define the Mellin moments for the polarized structure function $g_1^p(x,Q^2)$

$$g_1^p(N,Q^2) = \int_0^1 x^{N-1} g_1^p(x,Q^2)dx\,. \tag{1}$$

For the analysis presented in this paper, only the three lightest quark flavors, $q = u,d,s$, are taken into account. The twist–2 contributions to the structure function $g_1(N,Q^2)$ can be represented in terms of the polarized parton densities and the coefficient functions ΔC_i^N in the Mellin-N space by [2]

$$g_1^p(N,Q^2) = \frac{1}{2}\sum_q e_q^2\{(1+\frac{\alpha_s}{2\pi}\Delta C_q^N)[\Delta q(N,Q^2)+\Delta\bar{q}(N,Q^2)] + \frac{\alpha_s}{2\pi}2\Delta C_g^N \Delta g(N,Q^2)\}\,, \tag{2}$$

in this equation the NLO running coupling constant is given by

$$\alpha_s(Q^2) \cong \frac{1}{b\log\frac{Q^2}{\Lambda^2_{\overline{MS}}}} - \frac{b'}{b^3}\frac{\ln\left(\ln\frac{Q^2}{\Lambda^2_{\overline{MS}}}\right)}{\left(\ln\frac{Q^2}{\Lambda^2_{\overline{MS}}}\right)^2}\,, \tag{3}$$

where $b = \frac{33-2f}{12\pi}$ and $b' = \frac{153-19f}{24\pi^2}$. In the above equations, we choose $Q_0^2 = 1$ GeV^2 as a fixed parameter and Λ is an unknown parameter which can be obtained by fitting the experimental data. In Eq. (2), $\Delta q(N,Q^2) = \Delta q_v(N,Q^2) + \Delta\bar{q}(N,Q^2)$, $\Delta\bar{q}(N,Q^2)$ and $\Delta g(N,Q^2)$ are moments of the polarized parton distributions in a proton. ΔC_q^N, ΔC_g^N are also the N-th moments of spin-dependent Wilson coefficients given in [2].

According to improved polarized valon model framework, determination of the moments of parton distributions in a proton can be done strictly through the moments of the polarized valon distributions in Jacobi polynomials approach [10]. One of the simplest and fastest possibilities in the polarized structure function reconstruction from the QCD predictions for its Mellin moments is Jacobi polynomials expansion. According to [10] we have

$$xg_1^{p\,(N_{max})}(x,Q^2) = x^\beta(1-x)^\alpha \sum_{n=0}^{N_{max}} \Theta_n^{\alpha,\beta}(x)\sum_{j=0}^{n} c_j^{(n)}(\alpha,\beta) g_1^p(j+2,Q^2), \tag{4}$$

In our analysis we use $N_{max} = 9$, $\alpha = 3.0$ and $\beta = 0.5$. The same method has been applied to calculate the non-singlet structure function xF_3 from their moments [11]. Obviously the Q^2-dependence of the polarized structure function is defined by the Q^2-dependence of the moments.

THE PROCEDURE OF THE QCD FIT OF g_1^p DATA

For the QCD analysis presented in the present paper the following data sets are used: the HERMES proton data [9, 12], the SMC proton data [13], the E143 proton data [14], the EMC proton data [15, 16]. The number of the published data points above $Q^2 = 1.0$ GeV^2 for the different data sets are summarized in Table 1 of Ref. [10] for data on g_1 together with the $x-$ and Q^2–ranges for different experiments.

Using the CERN subroutine MINUIT [17], we defined a global χ^2 for all the experimental data points and found an acceptable fit with minimum $\chi^2/\text{d.o.f.} = 236.577/242 = 0.978$ in the LO case and $\chi^2/\text{d.o.f.} = 225.920/242 = 0.933$ in the NLO case. Our results for this analysis are reported in Ref. [10].

The flavor-asymmetry in the light-quark sea of the nucleon can be obtained from the contributions of unbroken sea quark distributions. We employ the polarized valon model and extract the flavor-broken light sea quark distributions with considering the Pauli-blocking ansatz[18, 19]. Same of this calculation was reported in [18] by using Bernstein polynomial method. Our results for δu, δd, $\delta\bar{u}$ and δd for different values of Q^2, using the Jacobi polynomials method are shown in Fig.(1).

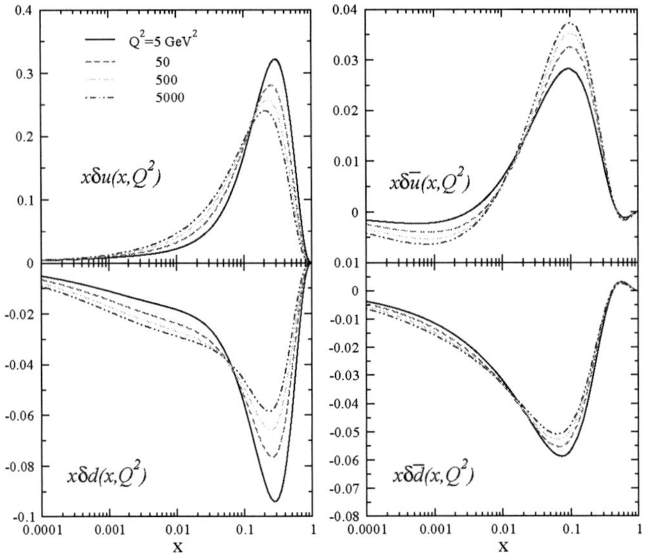

FIGURE 1. δu, δd, $\delta\bar{u}$ and $\delta\bar{d}$ as a function of x and for different values of Q^2.

By having polarized proton structure function, the deuteron structure function can be obtain. The deuteron structure function can be written as

$$g_1^d(x,Q^2) = \frac{1}{2}\,[g_1^p(x,Q^2) + g_1^n(x,Q^2)]\,(1 - \frac{3}{2}\omega_D)\,, \tag{5}$$

where $\omega_D (\simeq 0.058)$ accounts for the D-state admixture in the deuteron wave function [2]. Fig. (2) illustrates the deuteron data for $g_1^d(x,Q^2)$ at $Q^2 = 2,3,5$ GeV^2.

ACKNOWLEDGMENTS

We acknowledge Semnan university for financially supporting this project.

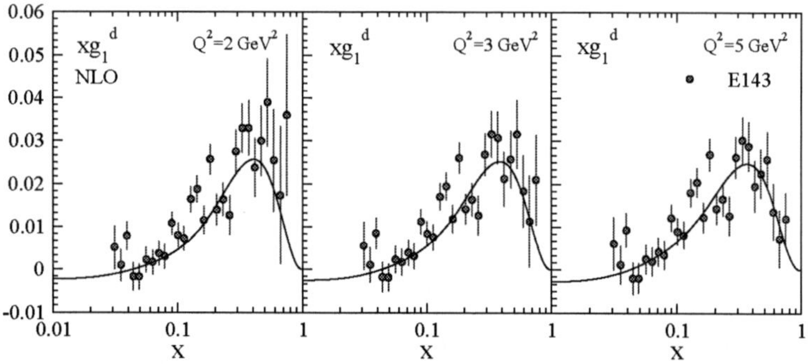

FIGURE 2. The polarized deuteron structure function g_1^d as a function of x and at $Q^2 = 2, 3, 5$ GeV^2.

REFERENCES

1. M. Anselmino, A. Efremov and E. Leader, Phys. Rept. **261** (1995) 1 [Erratum-ibid. **281** (1997) 399] [arXiv:hep-ph/9501369].
2. B. Lampe and E. Reya, Phys. Rept. **332** (2000) 1 [arXiv:hep-ph/9810270].
3. E. W. Hughes and R. Voss, Ann. Rev. Nucl. Part. Sci. **49** (1999) 303.
4. B. W. Filippone and X. D. Ji, Adv. Nucl. Phys. **26** (2001) 1, [arXiv:hep-ph/0101224].
5. A. N. Khorramian, A. Mirjalili and S. A. Tehrani, JHEP **0410** (2004) 062 [arXiv:hep-ph/0411390].
6. A. Airapetian *et al.* [HERMES Collaboration], Phys. Rev. D **71** (2005) 012003 [arXiv:hep-ex/0407032].
7. COMPASS collaboration (G. Baum et al.), CERN-SPSLC-96-14 (1996).
8. B. Adeva *et al.* [Spin Muon Collaboration], Phys. Lett. B **420** (1998) 180 [arXiv:hep-ex/9711008].
9. A. Airapetian *et al.* [HERMES Collaboration], Phys. Rev. D **75** (2007) 012007; [Erratum-ibid. D **76** (2007) 039901], [arXiv:hep-ex/0609039].
10. S. Atashbar Tehrani and A. N. Khorramian, JHEP **0707** (2007) 048.
11. A. L. Kataev, A. V. Kotikov, G. Parente and A. V. Sidorov, Phys. Lett. B **417**, (1998) 374 [arXiv:hep-ph/9706534].
12. A. Airapetian *et al.* [HERMES Collaboration], Phys. Lett. B **442** (1998) 484 [arXiv:hep-ex/9807015].
13. B. Adeva *et al.* [Spin Muon Collaboration], Phys. Rev. D **58** (1998) 112001.
14. K. Abe *et al.* [E143 collaboration], Phys. Rev. D **58**, 112003 (1998) [arXiv:hep-ph/9802357]; ; Airapetian *et al.*, Phys.Lett.B**442** (1998) 484.
15. J. Ashman *et al.* [European Muon Collaboration], Phys. Lett. B **206** (1988) 364.
16. J. Ashman *et al.* [European Muon Collaboration], Nucl. Phys. B **328** (1989) 1.
17. F. James, CERN Program Library Long Writeup D506; F. James and M. Roos, Comput. Phys. Commun. **10** (1975) 343.
18. A. Mirjalili, S. Atashbar Tehrani and A. N. Khorramian, Int. J. Mod. Phys. A **21** (2006) 4599 [arXiv:hep-ph/0608224].
19. A. N. Khorramian and S. Atashbar Tehrani, AIP Conf. Proc. **915** (2007) 502.

Shielding Calculation for High Energy Radiation X-Ray by Software program

Seyed Ali Rahimi

Department of Basic Sciences, Faculty of Health, Mazandaran University of Medical Sciences, Kilometer 18 KHAZARABAD Road,P.O.BOX:48175-1553,Tel:+981513543082,Fax:+981513542473-rahimi201@yahoo.com ,Sari-IRAN

Abstract. There are analytic methods for designing protective barriers however, they lack sufficient efficiency and considering the NCRP147(reports2007),designing mechanical protective barrier in order to protect the initial x-ray radiation and absorption of the ray quality of such radiation is different. In this study, computer software was designed to calculate the needed barrier with high accuracy. For proper determination of thickness of the protective barrier, relevant information about curves of radiation weakness, dose limit etc should be entered. The difference between the theoretical and calculated rates of this method is $X^2 = 10^{-5}$ which indicates accuracy and high efficiency of this software.

Keywords: Shielding, Diagnostic x-ray, Attenuation Curve, Building Materials
PACS: 07.05.-t , 01.50.H, 07.05.Fb, 07.05.Bx, 07.05.Dz, 07.05.Tp

INTRODUCTION

The following methods reassess shielding calculations in X-ray areas with respect to the methodology of the calculation of the barrier thickness and the number of sources consider in the area. Thus, they generate an overall solution for the cases met at the medical radiation structural design (1,2).This report provides an extension of an existing method for the calculation of the barrier thickness required to reduce the three types of radiation exposure emitted from the source, the primary, secondary and leakage radiation, to a specified weekly design limit (MPD). Because each of these three types of radiation are of different beam quality, having different shielding requirements, NCRP147 has provided means to calculate the necessary protective barrier thickness for each type of radiation individually(3,4).However, this report (NCRP147) provides little guidance for the contribution of each of the three types of radiation to the barrier thickness requirement.

CALCULATION OF THE BARRIER THICKNESS: A NEW METHOD

The medical physicist has to estimate which components of the field are the most important to be shielded and how they are to be combined; if more than one

CP1006, *High Energy Physics and Applications*, edited by J. Ellis, S. Nasri, and E. Malkawi

component is significant to generate a single shielding requirement. In questionable situations, multiple half-value layers (HVLs) of material is recommended to be added; by the general "add one half value layer (HVL)" approximation of NCRP147. Since the specified half value layers are those that are measured at high attenuation, he resultant barrier may be unnecessarily thick(5,6).Additionally, barrier requirements that are specified by using the techniques stated at NCRP147, will show enormous variations among users. Part of the variations is due to different assumptions made regarding the use of the examined room and the characteristics of adjoining space. Many of the differences, result from the difficulty of accurately relating information from the calculations to graphs and tables that are involved in the calculation process specified by this report. Moreover, the latest technological developments such as mammography, are not addressed and attenuation data for three-phase generators, that are most widely used today, is not provided (7,8).The design and shielding barriers in diagnostic X-ray departments, generally follows the ALARA principle. This is meant that, in practice, the exposure levels are kept "as low as reasonably achievable", taking into account economical and technical factors. Additionally, the calculation of barrier requirements includes many uncertainties (e.g. the workload, the actual kVp used etc.)(9,10).Figure 1 shows the excellent fit to the original curves obtained with the model. For a given value of K, no value of x found to differ by more than 0.03 mm from the original data.

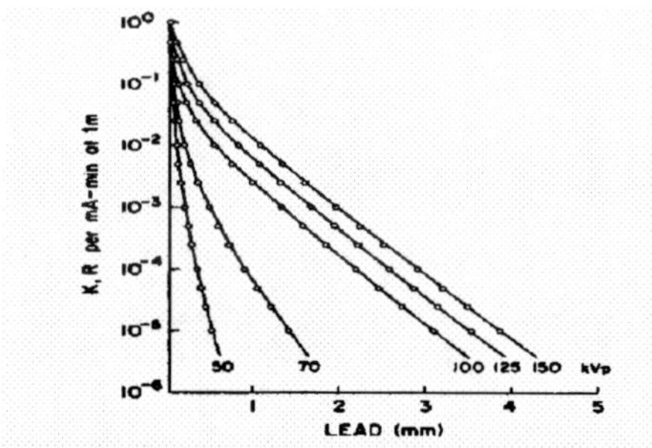

FIGURE 1.Values generated by the mathematical model of equation (1) (open circles) closely approximates to the attenuation curves from NCRP 147.

THE CALCULATION

Some of the variations among shielding requirements determined by various users are resulted from the difficulty of accurately relating information from the calculations, graphs and tables involved in the computation process specified by the NCRP147. In order to achieve simplification, Archer et al proposed a three-parameter model that accurately fit the published transmission data for lead in NCRP147. He also described an approach to account for the contribution of both scatter and leakage radiation. Each attenuation curve from NCRP147 Appendix D can be described by a curvilinear function, which increases rapidly at small values of lead shielding and then becomes horizontal at the large thickness. Such behavior was found by Archer et al to be well presented by a standard growth curve, which was integrated to obtain a mathematical representation (1):

$$K = K_0\left[\left(1+\beta/\alpha\right)e^{+\alpha\gamma\chi}-\beta/\alpha\right]^{-1/\gamma} \quad (1)$$

i) Where K is the number of Roentgens per mA-min per week at 1 m, K0 is the value of K with no lead in the beam, X is the thickness of lead in mm and α, β, γ were determined by the use of a modified non-linear least square program. Application of the model to primary barrier calculations is straightforward since if equation is set equal to the model, the resulting expression can be solved for the thickness of lead that is required to reduce the weekly exposure at Dpri to the maximum permissible exposure. Simplification of the NCRP 147 methods for determining secondary barriers is more difficult since both leakage and scattered radiation must be considered. The scattered radiation barrier is calculated from the equation described at NCRP 147 model. Therefore, setting this equation equal to that of (1), the required barrier thickness to protect for scatter radiation can be calculated. For leakage radiation and in order to simplify the procedure for computer calculation, the required number of HVLs can be related to the transmission factor BLx from the equation(2):

$$BLx = \frac{Pd_{sec}^2 \; 600I}{WT} \quad (2)$$

ii) The leakage barrier thickness can then be determined (3):

$$S_L = N \cdot HVL = \frac{-\ln\left\{\dfrac{Pd_{sec}^2 \cdot 600\, I}{WT}\right\}}{\ln 2} HVL$$

iii) The model and the simplification provided by equation (3) could be used to relate the thickness of barriers for leakage and scattered radiation to dsec. The actual exposure from leakage and scattered radiation at any value of dsec can also be determined for a specified barrier. This relationship is essential in determining the“exact”secondary barrier thickness required to meet the design limits. The total exposure at a point of interest, Ptot, is found by solving equations NCRP and (3) for P and adding them:

$$Ptot = PS + PL \quad (4)$$

iv) Where PL and PS are the contributions from leakage and scatter radiation respectively.Substitution yields(5):

$$P_{tot} = \frac{WT}{d_{sec}^{e}} \left[\frac{\exp\left(- \frac{Ln\ 2.x}{HVL} \right)}{600\ I} + \frac{ak_{ux}}{d_{sca}^{2} (400\ /F\ )} \right]$$

The above equation is then solved, to compute the appropriate thickness of material, to make the sum of the calculated leakage and scatter exposures equal to the weekly exposure limit.

v) Shielding Requirements for a Radiographic and a Chest Radiographic Unit in combination within the same room.

TABLE 1. I=4 mA,125 kVp, 1000 mA min/week Radio-graphic Examinations.

Barrier	P (mSv/week)	Use factor, U	Dpri (m)	Dsec (m)	Dleak (m)
Floor	0.02	1	2.0	1.2	2.0
Ceiling	0.02	0	0.0	1.8	1.0
Barrier 1	0.02	0.5	2.1	1.3	2.0
Barrier 2	0.12	0	0.0	3.0	3.8

TABLE 2. Chest Radio-graphic Examinations I=4 mA,125 kVp, 1000 mA min/week

Barrier	P (mSv/week)	Use factor, U	Dpri (m)	Dsec (m)	Dleak (m)
Floor	0.02	0	0.0	2.0	2.0
Ceiling	0.02	0	0.0	1.0	1.0
Barrier 1	0.02	1	2.0	2.5	3.0
Barrier 2	0.12	0	0.0	3.0	4.0

TABLE 3. Comparison results between these calculated by NCRP 147 and that described in this proposal.

	LEAD(mm)	Proposed method(mm)			CONCRETE	Proposed method(mm)		
Barrier	NCRP147	Table Recp Grid&cassette			NCRP147 (mm)	Table Recp Grid&cassette		
Floor	3.0	3.88	3.63	3.07	260.0	280	257	208
Ceiling	2.1	2.43	-	-	155.6	163	-	-
Barrier 1	2.7	3.53	3.30	2.77	240.0	258	235	191
Barrier 2	0.9	.811	-	-	66.8	67.6	-	-

NCRP147 METHODOLOGY

One of the advantages of the model is that it provides a concise representation of many different attenuation Curves(11,12). This greatly simplifies the task of designing a computer program to determine shielding requirements. Excellent agreement was found in the shielding requirements for primary barriers calculated by using the above mathematical three-parameter model and NCRP1 47 methodology(13,14).Moreover, as it has been described analytically, the presented method allows greater accuracy in the computation of secondary barrier(13,14). This is due to the fact that the shielding thickness required to reduce the weekly exposure to the design limit, can be precisely determined with no use of the “Add one HVL” recommendation of NCRP 147, which results in over shielding (15).

OUTLOOK

In this study, proper determination of required protective thickness for weakening of radiation at the permissible level is proposed. The multi radiation sources in a single room are considered and non guaranteed radiation of NCRP147 hypothesis is removed. The difference between the theoretical and calculated rates of this method is $X^2 = 10^{-5}$ which indicates accuracy and high efficiency of this software.

ACKNOWLEDGMENTS

Vice –chancellor of Mazandaran Medical Science University who undertake this research expenses with grant.

REFERENCES

1. A.A Okunade, L.A. Hussein "*An empirical model for x-ray attenuation in pb and mild steel*" Health physics Vol. 56, (June), pp.943-946, 1989.
2. Edward L. Nickoloff and Eileen M. Donnelly "*Use of gypsum drywall as shielding material for mammography*" Health physics Vol.54, No.6,(June) , pp. 465-468,1989.
3. C. Maccia, M Benedittini and C. Lefaure, F. fagnani "*Doses to patients from diagnostic radiology in France*" Health physics Vol.54, No.4, (April), pp. 397-340, 1988
4. Stewart C.Bushong and Sharon A .gGaze "*Radio-graphic workload and use factors for orthopedic facilities*" Health physics Vol.44, NO.1, (January). Pp.53-59, 1983
5. Benjamin R. Archer, T.R. Fewell ,B. J. Conway and P.W. Quinn "*Attenuation properties of diagnostic x-ray shielding materials*" Health physics Vol. 21, No.9. September, pp, 1499-1505, 1994
6. Almen A, Tingberg A, Mathson S, etal. *Influence of different technique factor s on image quality of lumbar spine radiographs as evaluated by establisher CEC image criteria.* Br J Radiol. 2000;73(875):1192-1199.
7. Muhogora WE, Nyanda AM, Kazema RR. *Experiences with the European guidelines on quality criteriafor radiographic images in Tanzania.* J Appl Clin Med Phys. 2001;2(4):219- 226.
8. Archer BR, Wagner LK. *Protecting patients by training physicians in flouroscopy radiation management.* J Appl Clin Med Phys. 2000.1(1):32-37.
9. Krutz C, Czapp W, Jrampe I, Leppek R, Klose KJ. *X-Ray taken by radiologists influence on a continous quality improvment process.* Rojo Fortschr Geb Rontgenstr Neuen Bildgeb Verfahr.2000; 172(4):391-396.
10. Kuon E, Schmitt M, Dahm JB. *Significant reduction of radiation exposure to operator and staff during cardiac interventions by analysis of radiation leakage and improved lead shielding.* Am J Cardiol. 2000;89(1):44-49.
11. Gogos KA, Yakoumakis EN, Tsalafoutus lA, Makri TK.*Radiation dose consideration in common peadiatric X-ray examinatio*n. Pediatr Radiol. 2003;33(4):236-240.
12. Statkiewicz Sherer MA, Visconti PJ, Ritenour ER. *Radiation protection in medical radiography 4th. St Louis: Mosby;2002.*
13. Sohrabi M; Borhan Azad S. and Taghzadeh S. Quality Control & Quality Assurance in Mammography, Forth Iranian Medical Physics Congress. 1999; 120-31.
14. Hart D, Hillier MC, Wall BF. *Doses to patients from medical X-ray examinations in the UK-2000 Review.* National Radiation Protection Board (NRPB)-W14. Chilton, UK: NRPB, 2002.
15. Rahimi S A, *Study on the performance of recommended standards in the diagnostic radiology units of the hospitals affiliated to the Mazandaran University of Medical Scienc,*Journal of X-Ray Science and Technology - Amsterdam Ios press- 15(2007)57-63- Netherlands.

Author Index

A

B

C

D

E

F

G

H

J

K

M

N

P

R

S

T

U

Y